ASCE STANDARD ASCE/SEI 49-12

American Society of Civil Engineers

Wind Tunnel Testing for Buildings and Other Structures

ASCE/SEI 49-12

This document uses both the International System of Units (SI) and customary units.

ASCE
AMERICAN SOCIETY OF CIVIL ENGINEERS

SEI ASCE
STRUCTURAL ENGINEERING INSTITUTE

Published by the American Society of Civil Engineers

Library of Congress Cataloging-in-Publication Data

Wind tunnel testing for buildings and other structures / American Society of Civil Engineers.
p. cm.—(ASCE standard ; ASCE/SEI 49-12)
Includes bibliographical references and index.
ISBN 978-0-7844-1228-2 (pbk.)—ISBN 978-0-7844-7690-1 (ebook)
1. Wind tunnel testing. 2. Wind-pressure. 3. Buildings—Aerodynamics. 4. Structural analysis (Engineering) I. American Society of Civil Engineers.
TA654.5W575 2012
690′.21—dc23

2012009848

Published by American Society of Civil Engineers
1801 Alexander Bell Drive
Reston, Virginia 20191
www.asce.org/pubs

This standard was developed by a consensus standards development process which has been accredited by the American National Standards Institute (ANSI). Accreditation by ANSI, a voluntary accreditation body representing public and private sector standards development organizations in the United States and abroad, signifies that the standards development process used by ASCE has met the ANSI requirements for openness, balance, consensus, and due process.

While ASCE's process is designed to promote standards that reflect a fair and reasoned consensus among all interested participants, while preserving the public health, safety, and welfare that is paramount to its mission, it has not made an independent assessment of and does not warrant the accuracy, completeness, suitability, or utility of any information, apparatus, product, or process discussed herein. ASCE does not intend, nor should anyone interpret, ASCE's standards to replace the sound judgment of a competent professional, having knowledge and experience in the appropriate field(s) of practice, nor to substitute for the standard of care required of such professionals in interpreting and applying the contents of this standard.

ASCE has no authority to enforce compliance with its standards and does not undertake to certify products for compliance or to render any professional services to any person or entity.

ASCE disclaims any and all liability for any personal injury, property damage, financial loss, or other damages of any nature whatsoever, including without limitation any direct, indirect, special, exemplary, or consequential damages, resulting from any person's use of, or reliance on, this standard. Any individual who relies on this standard assumes full responsibility for such use.

ISBN 978-0-7844-1228-2 (paper)
ISBN 978-0-7844-7690-1 (e-book)
Manufactured in the United States of America.

18 17 16 15 14 13 12 1 2 3 4 5

STANDARDS

In 2006, the Board of Direction approved the revision to the ASCE Rules for Standards Committees to govern the writing and maintenance of standards developed by the Society. All such standards are developed by a consensus standards process managed by the Society's Codes and Standards Committee (CSC). The consensus process includes balloting by a balanced standards committee made up of Society members and nonmembers, balloting by the membership of the Society as a whole, and balloting by the public. All standards are updated or reaffirmed by the same process at intervals not exceeding five years.

The following standards have been issued:

ANSI/ASCE 1-82 N-725 Guideline for Design and Analysis of Nuclear Safety Related Earth Structures
ASCE/EWRI 2-06 Measurement of Oxygen Transfer in Clean Water
ANSI/ASCE 3-91 Standard for the Structural Design of Composite Slabs and ANSI/ASCE 9-91 Standard Practice for the Construction and Inspection of Composite Slabs
ASCE 4-98 Seismic Analysis of Safety-Related Nuclear Structures
Building Code Requirements for Masonry Structures (ACI 530-02/ASCE 5-02/TMS 402-02) and Specifications for Masonry Structures (ACI 530.1-02/ASCE 6-02/TMS 602-02)
ASCE/SEI 7-10 Minimum Design Loads for Buildings and Other Structures
SEI/ASCE 8-02 Standard Specification for the Design of Cold-Formed Stainless Steel Structural Members
ANSI/ASCE 9-91 listed with ASCE 3-91
ASCE 10-97 Design of Latticed Steel Transmission Structures
SEI/ASCE 11-99 Guideline for Structural Condition Assessment of Existing Buildings
ASCE/EWRI 12-05 Guideline for the Design of Urban Subsurface Drainage
ASCE/EWRI 13-05 Standard Guidelines for Installation of Urban Subsurface Drainage
ASCE/EWRI 14-05 Standard Guidelines for Operation and Maintenance of Urban Subsurface Drainage
ASCE 15-98 Standard Practice for Direct Design of Buried Precast Concrete Pipe Using Standard Installations (SIDD)
ASCE 16-95 Standard for Load Resistance Factor Design (LRFD) of Engineered Wood Construction
ASCE 17-96 Air-Supported Structures
ASCE 18-96 Standard Guidelines for In-Process Oxygen Transfer Testing
ASCE 19-10 Structural Applications of Steel Cables for Buildings
ASCE 20-96 Standard Guidelines for the Design and Installation of Pile Foundations
ANSI/ASCE/T&DI 21-05 Automated People Mover Standards—Part 1
ANSI/ASCE/T&DI 21.2-08 Automated People Mover Standards—Part 2
ANSI/ASCE/T&DI 21.3-08 Automated People Mover Standards—Part 3
ANSI/ASCE/T&DI 21.4-08 Automated People Mover Standards—Part 4
SEI/ASCE 23-97 Specification for Structural Steel Beams with Web Openings
ASCE/SEI 24-05 Flood Resistant Design and Construction
ASCE/SEI 25-06 Earthquake-Actuated Automatic Gas Shutoff Devices
ASCE 26-97 Standard Practice for Design of Buried Precast Concrete Box Sections
ASCE 27-00 Standard Practice for Direct Design of Precast Concrete Pipe for Jacking in Trenchless Construction
ASCE 28-00 Standard Practice for Direct Design of Precast Concrete Box Sections for Jacking in Trenchless Construction
ASCE/SEI/SFPE 29-05 Standard Calculation Methods for Structural Fire Protection
SEI/ASCE 30-00 Guideline for Condition Assessment of the Building Envelope
SEI/ASCE 31-03 Seismic Evaluation of Existing Buildings
SEI/ASCE 32-01 Design and Construction of Frost-Protected Shallow Foundations
EWRI/ASCE 33-09 Comprehensive Transboundary International Water Quality Management Agreement
EWRI/ASCE 34-01 Standard Guidelines for Artificial Recharge of Ground Water
EWRI/ASCE 35-01 Guidelines for Quality Assurance of Installed Fine-Pore Aeration Equipment
CI/ASCE 36-01 Standard Construction Guidelines for Microtunneling
SEI/ASCE 37-02 Design Loads on Structures during Construction
CI/ASCE 38-02 Standard Guideline for the Collection and Depiction of Existing Subsurface Utility Data
EWRI/ASCE 39-03 Standard Practice for the Design and Operation of Hail Suppression Projects
ASCE/EWRI 40-03 Regulated Riparian Model Water Code
ASCE/SEI 41-06 Seismic Rehabilitation of Existing Buildings
ASCE/EWRI 42-04 Standard Practice for the Design and Operation of Precipitation Enhancement Projects
ASCE/SEI 43-05 Seismic Design Criteria for Structures, Systems, and Components in Nuclear Facilities
ASCE/EWRI 44-05 Standard Practice for the Design and Operation of Supercooled Fog Dispersal Projects
ASCE/EWRI 45-05 Standard Guidelines for the Design of Urban Stormwater Systems
ASCE/EWRI 46-05 Standard Guidelines for the Installation of Urban Stormwater Systems
ASCE/EWRI 47-05 Standard Guidelines for the Operation and Maintenance of Urban Stormwater Systems
ASCE/SEI 48-11 Design of Steel Transmission Pole Structures
ASCE/SEI 49-12 Wind Tunnel Testing for Buildings and Other Structures
ASCE/EWRI 50-08 Standard Guideline for Fitting Saturated Hydraulic Conductivity Using Probability Density Functions
ASCE/EWRI 51-08 Standard Guideline for Calculating the Effective Saturated Hydraulic Conductivity
ASCE/SEI 52-10 Design of Fiberglass-Reinforced Plastic (FRP) Stacks

ASCE/G-I 53-10 Compaction Grouting Consensus Guide
ASCE/EWRI 54-10 Standard Guideline for Geostatistical Estimation and Block-Averaging of Homogeneous and Isotropic Saturated Hydraulic Conductivity
ASCE/SEI 55-10 Tensile Membrane Structures
ANSI/ASCE/EWRI 56-10 Guidelines for the Physical Security of Water Utilities
ANSI/ASCE/EWRI 57-10 Guidelines for the Physical Security of Wastewater/Stormwater Utilities
ASCE/T&DI/ICPI 58-10 Structural Design of Interlocking Concrete Pavement for Municipal Streets and Roadways
ASCE/SEI 59-11 Blast Protection of Buildings
ASCE/EWRI 60-12 Guideline for Development of Effective Water Sharing Agreements

FOREWORD

The material presented in this standard has been prepared in accordance with recognized engineering principles. This standard should not be used without first securing competent advice with respect to its suitability for any given application. The publication of the material contained herein is not intended as a representation or warranty on the part of the American Society of Civil Engineers, or of any other person named herein, that this information is suitable for any general or particular use or promises freedom from infringement of any patent or patents. Anyone making use of this information assumes all liability from such use.

ACKNOWLEDGMENTS

The American Society of Civil Engineers (ASCE) acknowledges the work of the Wind Tunnel Testing for Buildings and Other Structures Standards Committee of the Codes and Standards Activities Division of the Structural Engineering Institute. This group comprises individuals from many backgrounds representing consumers, producers, and general interest individuals; included were wind engineers, consulting engineers, professors, and an insurance company representative.

This standard was prepared through the consensus standards process by balloting in compliance with procedures of ASCE's Codes and Standards Committee. Those individuals who serve on the Wind Tunnel Testing Standards Committee are:

Mark Baker
Jack Cermak
Cesar Farell
Peter Irwin
Nicholas Isyumov
Thusitha Kumarasena
Rusk Masih
Jon Peterka
Gene Sanders
Michael Soligo
Theodore Stathopoulos
Steven Stroh
Peter Vickery
ChongYul Yoon
Mehdi Zarghamee

CONTENTS

Chapter 1
GENERAL

1.1 SCOPE

This standard provides minimum requirements for wind tunnel tests to determine wind loads on and responses of buildings and other structures. Loads considered in this standard are wind loads for main wind-force-resisting systems (MWFRSs) and for individual structural components and cladding (C&C) of buildings and other structures. Loads produced by these tests are suitable for use in building codes and standards.

Provisions of this standard satisfy the requirements for wind tunnel testing of ASCE 7, *Minimum Design Loads for Buildings and Other Structures*. Wind tunnel testing has the capability to perform measurements beyond those specifically addressed in this standard, including pedestrian wind evaluations, dispersion of airborne pollutants, fugitive particulates, and wind energy siting studies. These studies are permitted to be included within the test report addressing wind loads.

1.2 REPORT CONTENT

The wind tunnel test report, hereinafter referred to as the report, shall be prepared to document the wind tunnel test and results. This report shall, at a minimum, include (1) a description of the properties of the simulated atmospheric boundary layer, including mean velocity and longitudinal turbulence intensity profiles, a spectrum of longitudinal turbulence with computed integral scale, and comparison to approved atmospheric models; (2) a description of the test and test methodology; and (3) results of the test. Detailed requirements are described in following sections.

The report shall address how the test laboratory met the requirements of this standard for instrumentation capability, and how the wind climate was defined and used to produce results given in the report.

For some wind tunnel tests, atmospheric boundary-layer modeling as described in this standard is not appropriate. Examples include (1) tests for parts and portions of a building where matching of turbulent length scales is not important for determining the important responses; and (2) tests that are intended to model nonboundary-layer winds. For these tests, the report shall discuss the deviation from the requirements of this standard and its implications, including the desirability of full-scale test validation.

Where wind tunnel test conditions vary from those specified in this standard, a justification for the variance shall be included in the report.

1.3 COORDINATE SYSTEMS

The following coordinate systems with mean and fluctuating velocities are defined:

x, y, z = right-hand coordinate axes, z is vertical;
U, V, W = mean velocities in x, y, z directions; and
u, v, w = fluctuating portion of velocity vector in x, y, z directions.

1.4 DEFINITIONS

Unless specified in this section, all terms used in this standard are consistent with definitions of the ASCE 7 wind load provisions.

Aeroelastic Feedback: The process by which motion of a body in the wind causes changes in applied wind pressures.

Aeroelastic Model: A model that simulates the stiffness, mass, and damping of the full-scale structure.

Approved: Acceptable to the authority having jurisdiction.

Atmospheric Boundary Layer (ABL): The lower part of the atmosphere, typically 800 to 2,000 ft (250 to 600 m) thick, in which the velocity increases with height and in which the flow is turbulent.

Atmospheric Surface Layer (ASL): The lower portion of the ABL in which the wind flow over flat, uniformly rough terrain has a logarithmic profile.

Integral (Macro) Scale: A characteristic length that represents a measure of the average size of turbulent eddies or gusts present in the atmospheric boundary layer.

Longitudinal Turbulence: Fluctuations in wind velocity in the direction of the mean wind.

Monte Carlo Simulation: A simulation process in which a series of events, such as hurricane speeds, is predicted by drawing on statistical properties of the underlying phenomena.

Neutral Thermal Stratification: Wind conditions in which thermal buoyancy forces are negligible; normally found in high-wind situations.

Nonboundary-Layer Winds: Winds that do not conform to the properties of the standard atmospheric boundary layer.

Recognized Literature: Published research findings and technical papers that are approved.

Reynolds Number: $U_b L_b / \nu$ (refer to the individual terms in Section 1.5, Symbols and Notation).

Turbulence: Random fluctuations in the wind velocity, or "gusts."

Upcrossing Procedure: A statistical method of calculating the mean recurrence rate for events of a stated magnitude.

Vortex Excitation: Time-varying pressures that are applied to a body predominantly in a direction perpendicular to the wind direction due to quasiperiodic shedding of vortices from alternate sides of the body.

1.5 SYMBOLS AND NOTATION

ABL = Atmospheric Boundary Layer;
ASL = Atmospheric Surface Layer;
h = mean roof height of a building or height of other structure;

L_b = characteristic length scale of a building or other structure;
${}^xL_u$ = integral length scale of longitudinal component of turbulence in the x direction;
n = power-law index for the variation of mean velocity with height;
u_* = surface shear velocity;
U = mean velocity at height z;
U_b = characteristic velocity scale of a building or other structure;
U_{ref} = mean velocity at height z_{ref};
U_s = mean velocity at height z_s;
z = height above ground;
z_0 = ground surface aerodynamic roughness length;
z_g = gradient height, the top of the atmospheric boundary layer;
z_{ref} = a reference height above ground where the mean velocity is measured or referenced;
z_s = height of the atmospheric surface layer, approximately 100 ft (30.5 m); and
ν = kinematic viscosity of air.

Subscripts

m = model;
g = gradient;
p = prototype (full-scale);
ref = reference; and
s = surface layer.

Chapter 2
SIMULATION OF WIND IN BOUNDARY-LAYER WIND TUNNELS

2.1 GENERAL

Conditions that the approach flow developed in a boundary-layer wind tunnel (BLWT) must satisfy for testing high-rise and low-rise structures are prescribed. Specifications are given for modeling the effects of nearby structures and topography on the approach flow. Simulation requirements herein are specified for the atmospheric boundary layer (ABL) with neutral thermal stratification (isothermal in the wind tunnel), which is appropriate for most tests to determine wind pressures and loads during strong winds. For aerodynamic stability tests of long-span bridge decks, where strong winds with stable stratification and low turbulence intensity can develop over water, tests in nonboundary-layer flows shall be permitted, provided that the rationale for the flow is discussed in the report. For testing of low-rise buildings and other structures, simulation of only the lower portion of the ABL or of the atmospheric surface layer (ASL) shall be permitted.

2.2 REQUIREMENTS FOR SIMULATION OF THE APPROACH FLOW

2.2.1 General. Approach flows that must be simulated are determined by roughness of the upwind fetch and height of the test structure. The reference approach flow is a turbulent boundary layer in which mean wind speed increases and turbulence intensity decreases with height.

2.2.2 The Reference Approach Flow. Dependence of the mean (averaged over 20–60 min, full scale) wind speed, $U(z)$, measured in wind tunnel simulations on upwind fetch roughness shall be in accordance with an ABL model published in approved literature. An example commonly used in wind engineering is the power-law profile:

$$\frac{U(z)}{U_{\text{ref}}} = \left(\frac{z}{z_{\text{ref}}}\right)^n \tag{2-1}$$

Profile parameters (power-law exponent n and gradient height z_g) for Eq. (2-1) are given in Table 2-1 for Exposures A, B, C, and D as defined in ASCE 7-93 (ASCE 1994). The mean wind speed power-law exponent for the simulated profile shall be within ±15% of the target value. Other profile exponents shall be permitted with justification provided in the report. In ASCE 7, n is referred to as $1/\alpha$, as discussed in the commentary.

TABLE 2-1. MEAN VELOCITY PROFILE PARAMETERS

Exposure[a]	n	z_g (m)
D	0.10	213
C	0.14	274
B	0.22	366
A	0.33	457

Note: n is the power law exponent.
[a]Exposure categories from ASCE 7.

For simulation of substantially the entire ABL, the reference quantities in Eq. (2-1) shall be set at a location in the wind tunnel to minimize interference from model structures. The relationship of the selected reference velocity and height to the approved reference boundary-layer model will be explained in the report. For modeling of low-rise buildings or smaller structures, simulation of only a portion of the ABL shall be permitted. For this case of partial ABL simulation, the relationship of the reference quantities U_{ref} and Z_{ref} to the approved boundary layer will be explained in the report. If the partial ABL simulation is deficient in similarity, as described in Section 2.2.3 (for example, missing spectral content at the low-frequency end of the spectrum), then any additional interpretation of the data shall refer to recognized literature for methods to make corrections.

Hurricane winds shall be simulated as prescribed for ABL winds in which U_{ref} is determined in accordance with provisions of Chapter 6. Selection of a C exposure for hurricane wind simulation over water approach shall be permitted.

2.2.3 Simulation of the Approach Flow

2.2.3.1 Conditions for similarity. A properly simulated ABL shall satisfy the following conditions:

1. The simulated mean velocity profile for a particular site and wind direction is consistent with the ABL profile or, in partial simulation for low-rise structures, a portion of the ABL or ASL.
2. Macro (integral) length scale, measured at the average height of the structure or other height related to test objectives of the longitudinal component of atmospheric turbulence, is simulated to be consistent with the scale of the modeled structure, and the turbulence intensity of this component of turbulence is consistent with that in the ABL or ASL at corresponding scaled heights.
3. To avoid excessive distortion of the flow, the geometric scale $(L_b)_m/(L_b)_p$ shall cause blockage no greater than 5% of the test section cross-sectional area for all wind directions unless specific measures are taken in the test procedures to account for the effects of blockage. The blockage area includes projected areas of both the test model and the proximity model on a plane normal to the test-section axis.
4. The surface-roughness Reynolds number for the upwind fetch, $u_* z_0/\nu$, satisfies Eq. (2-2).

$$\frac{u_* z_0}{\nu} \geq 2.5 \tag{2-2}$$

where u_* is the mean surface shear velocity and z_0 is the aerodynamic roughness length.

2.2.3.2 The proximity model. Modifications of the approach wind by buildings, structures, and local topographic features surrounding the model being tested in an ABL shall be simulated within a minimum radius of 820 ft (250 m). Common practice is to extend the proximity model over the entire turntable. This simulation shall be accomplished by placing simplified models (same scale as the test model) of the surrounding objects on the wind tunnel turntable at their proper locations. In the event a major object is located beyond the turntable that will significantly increase wind loading, a scaled model of the object shall be placed upwind of the turntable for appropriate wind directions during the testing program. Where simulation of a portion of the ABL or the ASL is used, the proximity model shall extend to the edge of the turntable. In addition, the effects of upwind influences shall be qualitatively evaluated to determine that modeling to a greater radius is not required.

2.2.3.3 Topographic models. Wind characteristics at a project site shall be established by a wind tunnel simulation for sites surrounded by extensive complex topography beyond the proximity model that result in approach flows not specified by Section 2.2.2 and for which local meteorological data do not exist. In such cases, a topographic model of small scale in the range of 1:500 to 1:5,000 shall be constructed and measurements made of mean wind speed and direction and turbulence at the site of study. Alternatively, a simpler test is permitted to obtain an upper bound for these flow properties, with the rationale justified in the report. The measurements shall be made for approach flows from a sufficient number of gradient wind directions to define flow over the site for all 360°, recognizing that the approach flow may be the same for some directions, and that satisfy the conditions set forth in Section 2.2.3.1. Model scales smaller than 1:5,000 shall be permitted provided the implications of the model scale are explained in the report.

2.3 GEOMETRIC SCALE CONSIDERATIONS

The geometric (length) scale selected for the building or other structure to be tested shall satisfy similarity criteria that ensure compatibility with lengths associated with the approach flow. Criteria that shall be satisfied for ABL and partial ABL simulations are as follows, in which subscripts m and p refer to model and prototype, respectively:

1. $$\left(\frac{L_b}{z_o}\right)_m = \left(\frac{L_b}{z_o}\right)_p \tag{2-3}$$

2. $$\left(\frac{L_b}{z_{\text{ref}}}\right)_m = \left(\frac{L_b}{z_{\text{ref}}}\right)_p \tag{2-4}$$

3. $$\left(\frac{{}^xL_u}{L_b}\right)_m = \left(\frac{{}^xL_u}{L_b}\right)_p \quad \text{or} \quad \left(\frac{{}^xL_u}{L_b}\right)_m \geq 3 \tag{2-5}$$

4. $$(L_bU_h/v)_m \geq 11{,}000 \tag{2-6}$$

The requirement of Eq. (2-3) shall be met within a factor of 2 to 3, reflecting the uncertainties in specification of Z_0 in the full scale. The requirement of Eq. (2-4) is readily met within 1% by selecting the identical reference locations for model and full scale. The first relationship in Eq. (2-5) shall be met within a factor of 3, and the second relationship should be adhered to as a minimum requirement. Eq. (2-6) is a minimum requirement. If any of Eqs. (2-3), (2-5), or (2-6) cannot be adhered to within these tolerances, the resulting uncertainties in the test results shall be discussed in the report, including any possible additional studies (such as high Reynolds number testing or tests in larger facilities) that may be needed to resolve the uncertainties.

Chapter 3
LOCAL AND AREA-AVERAGED WIND LOADS

3.1 GENERAL

Wind pressure measurements are carried out using rigid models equipped with pressure taps. *Local wind loads* refer to wind forces acting on small areas, such as individual windows. *Panel loads* refer to forces on larger portions of a structure, such as parts of walls supported by the structural frame or a portion of the roof supported by a single truss. The two load types are distinguished by the effects of area-averaging of wind pressure. The reference dynamic velocity pressure used in the wind tunnel shall be specified in the report of the study. In situations where no aeroelastic magnification occurs, and where Reynolds number effects are not significant, tests are permitted to be carried out only for a single wind tunnel speed. Instrumentation with frequency response sufficient to capture peak pressures of interest shall be used.

3.2 LOCAL PRESSURES

When measuring local exterior pressures to evaluate wind loads on components of the curtainwall and cladding systems, care shall be taken to ensure that measurements are obtained in all areas of significance to the designer, with particular attention to areas of high local pressure or potential aerodynamic "hot spots." Pressure shall be measured at a sufficient number of locations and wind directions so that no significant aerodynamic events are missed. Measurements of peak pressure magnitudes shall account for the statistical variability in individual samples of the peak.

3.3 AREA-AVERAGED PRESSURES

Area-averaged pressures are permitted to be evaluated by real-time spatial averaging of local pressures over the area of interest or by posttest processing of time series data on pressures measured simultaneously at multiple locations. The evaluation shall be carried out either by online algebraic addition in the data acquisition system of pressures measured simultaneously at a number of locations, or by pneumatic averaging using a single pressure transducer connected by a manifold to multiple taps. Measurement using porous surfaces and strain gauge elements shall be permitted.

For elements on which wind pressures can act on both sides (e.g., canopies, signs, solar collectors, parapets, freestanding walls), the differential pressures across the element shall be obtained by simultaneously measuring the area-averaged pressure on each side and algebraically summing the values. Direct force measurements on such elements are also permitted using specially designed instrumentation.

3.4 INTERNAL PRESSURES

Consideration shall be given to the effect of internal pressures and special care shall be taken in the evaluation of mean and dynamic internal pressures for different opening sizes, including the inherent porosity of the building envelope. For the assessment of both local and area-averaged loads, internal pressures shall be (1) evaluated analytically from the measured external pressures, (2) measured directly using appropriate wind tunnel models, or (3) obtained from a building code or standard.

Chapter 4
OVERALL WIND LOADS (EXCLUDING AEROELASTIC EFFECTS)

4.1 FORCE-BALANCE TECHNIQUE

For dynamically active structures that are expected to have insignificant aeroelastic response, the force-balance technique, or an equivalent procedure, is permitted. The high-frequency force-balance technique is used for overall wind-induced loads and responses of structures. The technique employs the use of a stiff balance/model system. Only the exterior geometry of the structure is reproduced. The force-balance system measures mean and fluctuating wind loads. The fluctuating base moments and base torsion represent the generalized modal forces for a structure having sway modes of vibration, which have deflection shapes that vary linearly with height, and a torsional mode of vibration, which has a deflection shape that is constant with height. The full-scale response of the structure, including the effects of dynamic load amplification, is determined by one of two possible approaches. The first is through the combination of the measured load spectra and dynamic properties of the structure using random vibration theory. The second is through a full time-history solution of the equations of motion of the structure in its lower modes of vibration, using the measured time-histories of generalized forces and torque as input. The force-balance shall be calibrated prior to wind tunnel testing. The calibration shall include measurements of sway and torsional fundamental frequencies of the vibration of the combined model/balance system.

A basic assumption of the force-balance approach is that the balance model system is rigid and thus the measured dynamic wind loads are not influenced by the dynamic properties of the balance/model system itself. The influence of any dynamic effects of the balance/model system shall be considered in the analysis methodology, and addressed in the report.

The effect of the departure of the true mode shapes of the full-scale structure from the linear and constant mode shapes inherently assumed in the force-balance methodology shall be considered and addressed in the report.

Aeroelastic tests (Chapter 5) are permitted as an alternative means of determining the response of structures that are dynamically active, but not necessarily aeroelastic.

4.2 INSTANTANEOUS SPATIAL PRESSURE INTEGRATION

The simultaneous measurement of pressures at multiple locations provides an alternative method of obtaining overall wind-induced loads and responses of structures. "Simultaneous" implies a rate of sampling significantly faster than the highest frequencies of interest in the load spectrum. The technique employs a real-time weighted integration of the exterior wind-induced pressures to produce estimates of modal forces acting on the structure. The full-scale response of the structure, including the effects of dynamic load amplification, is computed using the techniques used in the force-balance approach. This technique is relatively new, and caution should be exercised for buildings with complex geometry.

Chapter 5
AEROELASTICALLY ACTIVE STRUCTURES

5.1 GENERAL

Some flexible, lightweight, lightly damped structures experience aeroelastic feedback between the structure's motion and the wind forces. This can lead to unstable, wind-induced oscillations. Where experience or an initial assessment indicates that the aeroelastic amplification is sufficiently large to potentially affect the structure design, the wind tunnel tests shall include an examination of these effects using an appropriate aeroelastic model.

5.2 AEROELASTIC MODELING PROCEDURE

The aeroelastic model shall be capable of simulating the geometry and important structural responses of the full-scale structure using well-founded dynamic similarity principles. A full aeroelastic model of the entire structure, or alternatively, a partial aeroelastic model, such as a spring-mounted sectional model for a bridge, or a base-pivot model for a high-rise building, shall be constructed and tested.

The mass, stiffness, and damping of the structure, or ranges of these values, shall be scaled using accepted dynamic similarity principles. For partial aeroelastic models, the modes of vibration being simulated may or may not be identified, depending on use of the data, to permit use of rational analytical procedures for prediction of the full-scale structural behavior from the model data. Sufficient model scale shall be used to avoid inaccuracy caused by low Reynolds number effects.

The report shall include a description of the reasons for selecting the type of aeroelastic model, the model scaling principles, and the methodology and assumptions involved in extrapolating to full scale. The relevant full-scale and model mechanical properties, such as mass, polar moment of inertia, natural frequencies, and damping shall be described in the report. The treatment of the structural support conditions (e.g., fixed, pinned, sliding, foundation stiffness) and how they are simulated in the model shall be described. For full aeroelastic models, the frequencies and deflection shapes of the natural modes of vibration of the model shall be compared with those computed for the full-scale structure as an overall check on the model design and construction.

Where exact matching of significant scaling parameters is not achieved, the report shall discuss the reasons why an accurate match was not possible and what the probable effects on the wind tunnel results will be.

5.3 SCALING OF THE WIND

The mean velocity profile, turbulence intensities, and scales of the turbulence shall be matched as closely as possible to scaled properties of the natural wind at the site. Any deviations from the scale requirements as specified in Chapter 2 shall be assessed and discussed in the report.

Because some wind-induced oscillations, such as vortex excitation of long-span bridges, can be accentuated in wind with little turbulence, consideration shall be given to the range of possible turbulence levels, including low-turbulence conditions that might be present in stable atmospheric conditions.

5.4 TESTING

The aeroelastic model shall be instrumented in such a way that the relevant mean and dynamic responses can be recorded. The instrumentation shall have sufficient frequency response and signal-to-noise ratio to clearly identify the wind-induced response.

The range of wind speeds and directions examined in the tests shall be sufficient to cover the range of speeds and directions of importance to the structure, including angle of attack for bridges. This shall include speeds up to at least the basic wind speed used for structural strength design multiplied by the square root of the load factor. When testing for vortex excitation, due attention shall be paid to the possibility of excitation at low wind speeds.

Aeroelastic tests sometimes require low model-scale wind velocities, which may compromise accuracy due to Reynolds number effects. These effects shall be discussed in the report. The report shall describe the method used to measure the wind speed, including the location of the speed-measuring instrumentation in the wind tunnel.

Adequate precautions shall be taken to ensure that the model response is not adversely affected by spurious excitation from wind tunnel vibration or incident acoustic waves.

Chapter 6
EXTREME WIND CLIMATE

6.1 GENERAL

Where loads and/or building responses as a function of return period are a required project output, the wind tunnel data shall be combined with information on the local wind climate or other specified wind speed requirements. In regions not influenced by hurricanes, measurements of wind speeds recorded at nearby airport stations provide an acceptable source of wind speed data. In regions influenced by hurricanes, a Monte Carlo analysis for prediction of hurricane speeds is permitted.

6.2 ANALYSIS OF WIND DATA

A wind climate model is permitted to be derived using either an extreme-value approach or through the use of an upcrossing procedure. In either case, due regard shall be given to the length of the wind speed record (sampling error), anemometer height, data quality, averaging time, and terrain exposure. Statistical descriptions of the wind climate are permitted to be developed using upper-level (balloon) data in combination with an upcrossing approach to derive the wind climate. The report shall document the analysis methodology and wind-speed data source. A reduction in wind loads resulting from the incorporation of the impact of wind direction is permitted. Load reductions due to wind directional effects shall be limited to one-third of the result obtained without consideration of wind direction effects. Exceptions to the one-third limitation shall be permitted for structures located where only a limited range of wind directions is possible, such as for a structure at the bottom of a narrow valley.

6.3 HURRICANES

The influence of hurricane wind speeds on the wind climate and wind-induced loads and responses shall be considered in hurricane-prone regions. Wind loads used in the design of a structure or its components shall be determined using the combined effects of hurricane and nonhurricane winds.

6.4 OTHER WINDS

Where downslope winds occur, they shall be included in the analysis of structure loads and responses. Estimate of design downslope wind speed using available local knowledge is permitted where measured data are not available. Inclusion of the effects of tornadoes and microbursts on structure loads and responses is permitted.

6.5 COMBINATION OF WIND AND WIND TUNNEL DATA

Wind tunnel data shall be combined with wind climate data using approved statistical processes.

Chapter 7
SNOW LOAD MODEL STUDIES

The ASCE 7 snow load provisions are based primarily on full-scale observations of snow accumulations on simple, common shapes of roofs, including allowances for the nonuniform loads that can arise due to drifting and sliding. Scale-model studies in wind tunnels or water flumes are permitted to assist in interpretation of ASCE 7.

The methods that have been used to model snow drifting can be divided into two categories:

1. Those in which the consequences of particular storms are simulated using scaled models and introducing particles into the wind tunnel or water flume to simulate snow particles and their accumulations; and
2. Those in which the wind velocity patterns are measured on scaled models and the snow drifting and accumulation are numerically inferred with reliance on field data on snow transport rates and other information. Snow particles are not physically simulated in this approach.

Methods in both categories are useful for identifying the potential formation of unusual snowdrifts not accounted for by ASCE 7 due to individual storms and, in some cases, due to the effects of multiple storms. In some climatic regions, maximum snow loads result from the cumulative effects of many snowfalls over a prolonged part of the winter season. In such situations it becomes necessary to track snow accumulations on an hour-by-hour basis by recognizing the consequences of each snowfall and using numerical methods to keep track of snow drifting, snow melting, and the percolation and runoff of melt and rainwater. Both Category 1 and 2 methods can be used in combination with long-term meteorological records to make predictions of snow loads on a statistical basis, including the cumulative effects over extended periods of drifting, snowfall, melting, refreezing, and absorption of rain.

The report shall include a description of the model scaling principles, computational methods and assumptions used, and state the limitations of the study due to any mismatch of dynamic similarity parameters or other causes.

Chapter 8
ACCURACY, PRECISION, AND QUALITY ASSURANCE

The instrumentation used in wind tunnel model tests of wind effects shall be capable of providing adequate measures of the mean and, where necessary, the dynamic or time-varying response over periods of time corresponding to 1 to 2 h in full scale. For quantitative measurements of time-varying local pressures, overall wind loads, and aeroelastic responses, the frequency response of the instrumentation system shall be sufficiently high to permit meaningful measurements at all relevant frequencies without magnitude and phase distortions. Consideration of phase becomes important in spatial averaging procedures, where simultaneous measurements of local pressures are combined to form dynamic loads on particular tributary areas or on the entire building or structure.

All measurements shall be free of significant acoustic effects, electrical noise, mechanical vibrations, and spurious pressure fluctuations, including fluctuations of the ambient pressure within the wind tunnel caused by the operation of the fan, door openings, and the influence of atmospheric wind. Where necessary, corrections shall be made for temperature drift.

The reliability of all wind tunnel data shall be addressed, including both the accuracy of the overall simulation and the accuracy and the repeatability of the measurements. Checks shall be devised where appropriate to ensure the reliability of the results. These shall include basic checkout procedures of the instrumentation, including its traceable calibration, the repeatability of individual measurements, and, where feasible within cost constraints, comparisons with similar data obtained by different methods. For example, mean overall force or moment can be compared with similar data obtained from the integration of mean local pressures. Credibility checks with codes, full-scale data, and existing experience shall be included where such comparisons are available and meaningful. In order to avoid multiple interpretations, comparison with wind load codes or standards shall not be included in the report unless specifically requested by the sponsor.

Comparisons with full-scale data are important to ensure that experimental techniques and analysis procedures provide representative information. Such comparisons have significant difficulties and are beyond the scope of most measurement cases. Where model and full-scale comparisons are performed, attention shall be paid to the fact that processes are stochastic, and allowance shall be made for their inherent variability.

COMMENTARY

This commentary is not a part of the ASCE Standard 49-12, *Wind Tunnel Testing for Buildings and Other Structures*. It is included for informational purposes.

This commentary consists of explanatory and supplementary material designed to assist engineers, local building code committees, and regulatory authorities in applying the recommended requirements. In some cases it will be necessary to adjust specific values in the standard to building-specific conditions. In others, a considerable amount of detailed information is needed to put the provisions into effect. This commentary provides a place for supplying material that can be used in these situations and is intended to create a better understanding of the recommended requirements through brief explanations of the reasoning employed in arriving at them.

The sections of the commentary are numbered to correspond to the sections of the standard to which they refer. Because it is not necessary to have commentary material for every section in the standard, there are gaps in the numbering in the commentary.

Chapter C1
GENERAL

C1.4 DEFINITIONS

The following additional definitions are used herein:

Cauchy Number: $E / (\pi U_b^2)$;
Froude Number: $U_b^2 / (g L_b)$;
FUR: flat uniformly rough terrain; and
PBL: planetary boundary layer.

C1.5 SYMBOLS AND NOTATION

The following additional symbols and notation are used herein:

E = Young's modulus;
f_c = Coriolis parameter = 1×10^{-5} rad/s;
Hz = hertz, cycles per second;
U_{geo} = mean velocity at geostrophic height;
z_{geo} = geostrophic height;
κ = 0.4;
n = frequency; also power-law index for the variation of mean velocity with height;
N = a reference frequency;
ρ = density of air;
φ = latitude;
ω = angular velocity of the Earth; and
°C = degrees Celsius.

Chapter C2

SIMULATION OF WIND IN BOUNDARY-LAYER WIND TUNNELS

C2.1 GENERAL

Modeling of wind in atmospheric boundary layers (ABLs) with neutral thermal stability is appropriate for wind tunnel tests to determine wind effects on buildings and other structures. For this application, strong winds [speeds greater than 33 ft/s (10 m/s) at height of 33 ft (10 m)] are of concern in which mixing by mechanically generated turbulence destroys thermal stratification. Temperature variation with height in the resulting neutral flow is −5.4°F/1,000 ft (−1°C/100 m) (adiabatic lapse rate) for the atmosphere and uniform temperature (isothermal flow) for the wind tunnel. Simulation criteria for the wind tunnel flow are based on wind characteristics for neutral atmospheric flow over flat uniformly rough (FUR) terrain. Details of techniques to achieve these simulations and their limitations are discussed in ASCE MOP 67, *Wind Tunnel Studies of Buildings and Structures* (ASCE 1999).

An example of a mean velocity profile throughout the planetary boundary layer (PBL) in the absence of thermal effects is shown in Fig. C2-1 (Lettau 1962). This type of velocity profile has been confirmed by measurements over Jacksonville, Florida (Holton 1971). For different values of z_0, U_{geo}, and thermal conditions, the profile shape varies from that shown in Fig. C2-1.

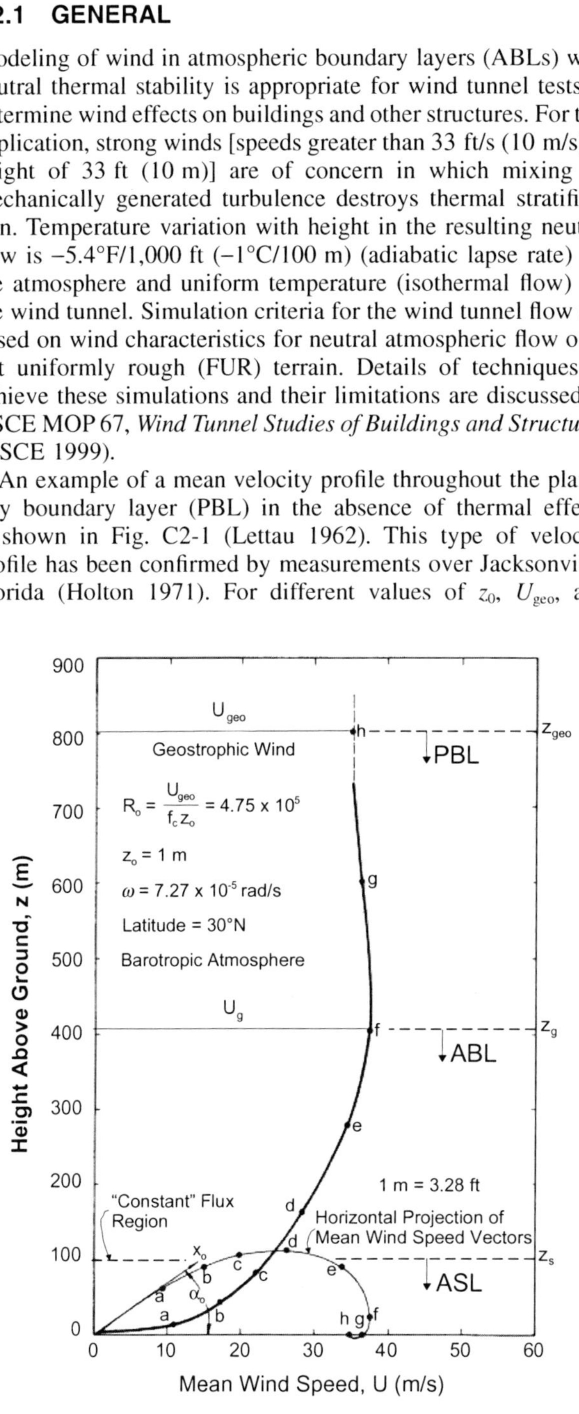

FIG. C2-1. EXAMPLE OF A PBL CALCULATED BY CERMAK (1975) USING MODEL OF LETTAU (1962).

C2.2 REQUIREMENTS FOR SIMULATION OF THE APPROACH FLOW

C2.2.1 General. For wind engineering purposes, the ABL height is defined as the height where the mean wind speed, U, ceases to be affected by shear stress at the surface. This height and the corresponding U are designated as gradient values z_g and U_g, respectively. The height of the PBL, where forces caused by the Coriolis acceleration and pressure gradient are in equilibrium, is designated as the geostrophic level, z_{geo}. At the geostrophic level, mean wind speed is designated by U_{geo}. The lower portion of the ABL, where turbulent fluxes vary within about 10% of surface values, is defined as the atmospheric surface layer (ASL) with a height, z_s. For middle-latitude and neutral flow over FUR terrain, Panofsky and Dutton (1983) give the following expression for estimation of z_{geo}:

$$z_{geo} = 0.175\ u_*/f_c \qquad \text{(C2-1)}$$

where the Coriolis parameter $f_c = 2\,\varpi\,\sin\Omega$, ϖ is angular velocity of the Earth (7.27×10^{-5} rad/s), and Ω is the latitude. Eq. (C2-1) ceases to be valid near the equator, where f_c approaches zero. An approximate value for z_s is 0.1 z_{geo}.

C2.2.2 The Reference Approach Flow. An empirical formulation for the mean velocity profile throughout the ABL, except for very near the surface, is given by the "power law." This relationship, proposed by Hellman (1916) and quantified by Davenport (1960), is given by

$$U(z)\ /\ U(z_{ref}) = (z/z_{ref})^n \qquad \text{(C2-2)}$$

or, when the reference height $z_{ref} = z_g$,

$$U(z)\ /\ U_g = (z/z_g)^n \qquad \text{(C2-3)}$$

The power law exponent, n, as well as z_g, vary with the surface roughness length, z_0. Values for a range of FUR terrain types perturbed by scattered obstacles, along with corresponding exposure classifications used in ASCE 7, are listed in Table C2-1.

An expression for mean velocity profiles within the entire PBL proposed by Harris and Deaves (1980) is given by the following equation:

$$U(z)/u_* = \frac{1}{\kappa}[\ln(z/z_0) + a_1\zeta + a_2\zeta^2 + a_3\zeta^3 + a_4\zeta^4] \qquad \text{(C2-4)}$$

where

$\zeta = z/z_{geo}$;
$a_1 = 5.75$;
$a_2 = 1.87$;
$a_3 = -1.33$; and
$a_4 = 0.25$.

TABLE C2-1. NEUTRAL BOUNDARY-LAYER PARAMETERS

Class	Terrain Description	$(z_0)_r$[a] (m)	n[b]	$1/\alpha$[c]	I_u[d] (%)	${}^xL_u$[e] (m)	Exposure[f]	z_g[f] (m)
1	Open sea, fetch at least 3 mi (5 km)	~0.0002	0.10	0.09	9.2	190	D	213
2	Mud flats, snow; no vegetation, no obstacles	0.005	0.13		13.2	140	—	—
3	Open flat terrain; grass, few isolated obstacles	0.03	0.14	0.11	17.2	110	C	274
4	Low crops; occasional large obstacles, $X'/h > 20$	0.10	0.18		21.7	84	—	—
5	High crops; scattered obstacles, $15 < x'/h < 20$	0.25	0.22	0.14	27.1	64	B	366
6	Parkland, bushes; numerous obstacles, $x'/h \sim 10$	0.5	0.29		33.4	55	—	—
7	Regular large-obstacle coverage (suburb, forest)	1.0–2.0	0.33	0.20	43.4	45	A	457
8	City center with high- and low-rise buildings	>2	0.40–0.67		—	—	—	—

Note: x' is typical spacing between obstacles; h is the height of a typical structure or obstacle.
[a]Regional roughness lengths from Wieringa (1992).
[b]Mean velocity power-law exponents from Davenport (1960) and ASCE 7-93 (ASCE 1994).
[c]3-s gust power-law exponents from ASCE 7-95 (ASCE 1996), ASCE 7-98 (ASCE 2000), and ASCE 7-02 (ASCE 2003).
[d]Turbulence intensities for FUR terrain (z = 10 m) with the same "local" roughness length according to Eq. (C2-6).
[e]Integral lengths of turbulence for U_{10} = 20 m/s, z = 10 m, and $f_c = 1 \times 10^{-4}$ rad/s from *ESDU 85020* (ESDU 1993).
[f]Exposure categories and gradient heights from ASCE 7.

This formulation presumes Eq. (C2-1) to be valid—an assumption that is reasonable for mid-latitudes, but has not been shown to apply at low latitudes.

Similarity of the mean velocity profile, $U(z)$, is achieved when the exponent n in Eq. (C2-3) corresponds to n in Table C2-1 for terrain appropriate for the site. Because of uncertainties in classification of FUR terrain, velocity profiles realized with n differing from values given in Table C2-1 for Exposures A, B, C, and D by less than ±15% are acceptable simulations. In cases where the actual exposure is between the ASCE exposures or where exposure varies with wind direction, the wind tunnel test should use the most appropriate boundary layer that applies for each approach wind direction.

Mean velocity profiles in the ASL are given by the well-established logarithmic distribution

$$U(z) = (u_*/\kappa)\ln(z/z_0) \quad \text{(C2-5)}$$

where κ, the von Kármán constant, = 0.4, and z_0 = the aerodynamic surface roughness length [the height where $U(z_0) = 0$]. The mean friction velocity, u_*, is defined as $(\tau_o/\rho)^{1/2}$ where τ_o = the mean surface shear stress, and ρ = the mass density of air. Eq. (C2-5) is used in practice to obtain values of u_* and z_0 for a particular location by making simultaneous measurements of U at three or more heights in the ASL.

Turbulence characteristics in the ABL for neutral FUR conditions are strongly dependent upon the surface roughness. Turbulent flow velocities vary randomly in time and space; therefore, statistical descriptions are necessary. Statistics of primary interest for wind engineering applications to wind effects on buildings and structures are variances or standard deviations, σ_i, power spectral densities, $S_i(z, n)$, and integral scales, ${}^jL_i$, for the ith velocity component and the jth coordinate direction.

Turbulence intensities $I_i(z) = \sigma_i(z)/U(z)$ in the ASL have been studied extensively by means of micrometeorological measurements (Kaimal et al. 1972; Busch 1973; Panofsky and Dutton 1983). As a result, the turbulence intensities can be expressed as

$$I_i = A_i / \ln(z/z_0) \quad \text{(C2-6)}$$

where $A_i = 0.4\,(\sigma_i/u_*)$ = 1.0, 0.8, and 0.5 for $i = u$, v, and w, respectively. Values of I_u at z = 10 m are given in Table C2-1 for various terrain types.

The power spectral density for the longitudinal component (direction of mean velocity) of turbulence, $S_u(z)$, has been investigated more thoroughly than for the other two components. In an ideal simulation of the ABL, $S_u(z)_m$ for the wind tunnel approach flow should match $S_u(z)_p$. Widely used equations for power spectra were developed by von Kármán (1948) based on the theory of isotropic turbulence. Equations for the u and w components are as follows:

$$\frac{nS_u(z,n)}{\sigma_u^2} = \frac{4n\,{}^xL_u/U}{[1+70.8(n\,{}^xL_u/U)^2]^{5/6}} \quad \text{(C2-7)}$$

and

$$\frac{nS_w(z,n)}{\sigma_w^2} = \frac{4n\,{}^xL_w/U[1+755.2(n\,{}^xL_w/U)^2]}{[1+283.1(n\,{}^xL_w/U)^2]^{11/6}} \quad \text{(C2-8)}$$

where ${}^xL_w$ is the integral scale of the vertical component of velocity in the x direction. Irwin (1979b) has developed equations for cross-spectra of u and w in terms of separation distance.

Using a general nondimensional formulation of Olesen et al. (1984), Tieleman (1995) developed spectral equations for u, v, and w over ideal and slightly perturbed FUR terrains. These equations for the longitudinal component u are as follows:

Ideal FUR terrain

$$\frac{nS_u(z,n)}{\sigma_u^2} = \frac{20.53\,f}{1+475.1\,f^{5/3}} \quad \text{(C2-9)}$$

Perturbed FUR terrain

$$\frac{nS_u(z,n)}{\sigma_u^2} = \frac{40.42\,f}{(1+60.62\,f)^{5/3}} \quad \text{(C2-10)}$$

where n = frequency in hertz and f = the reduced frequency (Monin similarity parameter) $nz/U(z)$. The spectrum proposed by Davenport (1961), which does not provide for height dependence, is given by

$$\frac{nS_u(n)}{\sigma_u^2} = \frac{0.64x^2}{(1+x^2)^{4/3}} \quad \text{(C2-11)}$$

where $x = 1{,}200n/U_{10}$ and U_{10} = the mean wind speed in m/s at z = 10 m. Extensive discussions on the basis for and differences of the various power spectral density formulations are available in published literature (Simiu 1974; Simiu and Scanlan 1996; Tieleman 1992; Kaimal and Finnigan 1994).

Nine integral scales are defined by ${}^jL_i$, where j is the separation coordinate (x, y, z) and i is the velocity component (u, v, w);

however, ${}^xL_u$ is the most common measure of the most energetic gust size used in wind engineering practice. Other components of ${}^jL_i$ are significant, but ${}^xL_u$ has been the most accessible from the usual wind speed measurements on an instrumented tower. A good estimate of ${}^xL_u$ within the ASL is given by the following equation:

$$ {}^xL_u = z/(2\pi f_m) \quad \text{(C2-12)} $$

where f_m = the reduced frequency for which $nS_u(n)$ is a maximum. The value of f_m for Eq. (C2-9), the ideal FUR terrain, = 0.032, and for Eq. (C2-10), the perturbed FUR terrain, f_m = 0.025. Analysis by Counihan (1975) of various sets of atmospheric turbulence data within the ASL indicates that ${}^xL_u$ decreases with increasing surface roughness and that ${}^yL_u \cong 0.3\ {}^xL_u$ and ${}^zL_u \cong 0.5\ {}^xL_u$. Data for ${}^xL_u$ given in *ESDU 85020* (ESDU 1993) for U_{10} = 20 m/s and $f_c = 1 \times 10^{-4}$ rad/s are listed for a height of 10 m and 200 m in Table C2.1. At z = 10 m, ${}^xL_u$ decreases with increasing z_0, but at 200 m, ${}^xL_u$ increases with increasing z_0.

C2.2.3 Simulation of the Approach Flow

C2.2.3.1 Conditions for similarity. Wind tunnels designed for simulation of the ABL are identified as boundary-layer wind tunnels (BLWTs). An array of vortex generators (Counihan 1969) or an array of spires (Standen 1972) is commonly inserted at the test-section entrance to increase the boundary-layer depth, δ, and the effective length of fetch, L_f (Cermak 1975, 1982). A flow-conditioning system of horizontal vanes can be inserted at the test-section entrance to augment depth of the ASL for studies of low-rise buildings (Cermak et al. 1993; ASCE 1999). Boundary-layer wind tunnels are configured in the form of either an open circuit or a closed circuit. Typical test-section characteristics of BLWTs that have been found to satisfy the requirements stipulated in Section 2.2.3.1 are

1. Approximate length ≥ 49 ft (15 m);
2. Approximate width ≥ 6.5 ft (2 m);
3. Approximate height ≥ 6.5 ft (2 m);
4. Maximum flow speed ≥ 29.5 ft/s (9.0 m/s); and
5. Adjustable ceiling height for control of the longitudinal pressure gradient, $\partial p/\partial x$.

Pressure variation in the mean flow direction is approximately zero for the ABL. However, if not corrected, pressure along the length of a constant cross-section test section with a rough floor and building and proximity models will decrease significantly and modify flow separation and reattachment on the model. The longitudinal pressure gradient, $\partial p/\partial x$, and other flow distortions will generally be acceptable if the blockage does not exceed 5%. Larger blockage is frequently permissible if special care is taken to control or limit blockage errors or if analytic/empirical blockage corrections can be shown to apply.

The relationships given by Eqs. (C2-2), (C2-3), and (C2-6) through (C2-12) will be independent of mean wind speed (Reynolds number-independent) if the surface roughness and wind speed are sufficiently large. Characteristics of the simulated ASL will be Reynolds number-independent if the criterion given by Eq. (2-2) is satisfied (Schlichting 1979). Eq. (C2-5) can be used to determine u_* for the wind tunnel flow. Additional information on details and performance of BLWTs is available in published literature (Cermak 1981, 1982; Davenport et al. 1985).

C2.2.3.2 The proximity model. Models of buildings and other objects that comprise the proximity model are commonly fabricated from wood or dense styrofoam. The building models may be simplified in geometry by replication of overall cross-section and height dimensions in block form. Street names should be identified by labels on the proximity model, along with a north-direction indicator that will be visible in photographs to be included in the report.

C2.2.3.3 Topographic models. Meteorological data for flow over complex topography reveal that mean velocity distributions and turbulence statistics (variances, integral scales, and spectra) are significantly different from that of flow over FUR terrain having the same surface roughness (Panofsky and Dutton 1983). Therefore, small-scale regional topographic models are needed to provide wind data at sites located in or downwind from complex topography. Examination of time series of wind direction in flow over complex terrain show 50° to 100° variations in short time spans of about 1 s (Tieleman 1992). Although walls of a BLWT tend to restrict lateral direction fluctuations, local flow characteristics measured over topographic models scaled in the range 1:500 to 1:5,000 compare well with corresponding field measurements for neutral thermal conditions (Cermak 1984). Studies by Meroney (1980) of flow over the Rakaia Gorge region of New Zealand using a 1:5,000 scale model gave wind speed and direction data that were in good agreement with field measurements. Model scales smaller than 1:5,000 are possible, but effects of molecular viscosity, atmospheric stability, and Coriolis acceleration play an increasing role and measurements near the model surface become more uncertain; the potential impact of these effects should be discussed in the report.

C2.3 GEOMETRIC SCALE CONSIDERATIONS

Choice of a model length scale is conditioned by dimensions of the BLWT test section and depth of the simulated ABL $(z_g)_m$ or the simulated ASL $(z_s)_m$ shown in Fig. C2.1. For wind tunnels with dimensions in the range given in Section C2.2.3.1, $(z_g)_m$ is commonly in the range of 2.5–5 ft (0.75–1.50 m). For studies of low-rise buildings and other structures, $(z_s)_m$ can be increased to depths in the range of 1.6–2.5 ft (0.5–0.75 m) (Cermak et al. 1993; ASCE 1999). The geometric scale $(L_b)_m/(L_b)_p$ must be selected to keep blockage within 5%, unless corrections are made, while satisfying criteria [Eqs. (2-3), (2-4), (2-5), and (2-6)] for similarity with the approach flow.

The prototype Jensen number (Jensen 1958) of Eq. (2-3) can be established by selecting the appropriate value of z_0 from Table C2.1 for terrain upwind of the site. For a particular site, the appropriate z_0 may be different for different wind directions. In practice, a common procedure is to adjust the upwind surface roughness until the desired power-law exponent, n, (Table C2.1) is achieved.

Choice of a model length scale is strongly dependent on proper scaling relative to macroscale of the simulated boundary-layer as expressed by Eq. (2-4). For modeling of the entire ABL, the reference length z_{ref} corresponds to z_g. For low-rise building studies, modeling of only the ASL is necessary and z_{ref} corresponds to z_s. In order to satisfy the Reynolds number criterion of Eq. (2-6), the model scale should be as large as possible (with due consideration of blockage limitations). Therefore, wind tunnel tests should be performed with the maximum value of z_g (or z_s) possible for a particular wind tunnel.

Approximate scaling with respect to the integral length scale of turbulence ${}^xL_u$ according to Eq. (2-5) is ordinarily satisfied when the previously discussed conditions have been satisfied. The criterion of Eq. (2-5) can be checked by using Eq. (C2-12) to estimate ${}^xL_u$ for the prototype and by measurement of $S_u(z,n)$ for the simulated flow.

Chapter C3
LOCAL AND AREA-AVERAGED WIND LOADS

C3.2 LOCAL PRESSURES

To avoid missing any significant aerodynamic effects for a major building, pressure measurements at some 800 or more locations may be made, depending on the complexity of the exterior geometry. Representative information for low-rise or less wind-sensitive buildings can be obtained with a more modest pressure tap coverage. Measurements shall generally be made at azimuth increments of no more than 10°. Mean values of local pressure are permitted to be integrated over exterior surfaces of a building or structure to determine the mean or static components of the overall wind loading. Taps in the model must be checked for correct connections and for leakage.

The model-to-full-scale frequency ratio is given by

$$n_m/n_p = (L_p/L_m)\ (U_m/U_p) \qquad \text{(C3-1)}$$

where n = frequency, L = a characteristic building length, U = velocity, and subscript m and p refer to model and prototype, respectively. For typical ratios of $(L_p/L_m) = 400$ and $(U_m/U_p) = 0.25$, model frequencies are about 100 times faster than full scale. Higher values of n_m permit higher effective measured values of n_p (Holmes 1982).

The largest peak local pressures have an effective duration of about 1 to 3 s (full-scale). This duration is long enough to cause a cladding element with a natural frequency in excess of several hertz to respond fully to the load (quasistatic response). Therefore, maximum loads are caused by short-duration peak pressures acting with no dynamic magnification by the cladding element. However, the possibility of resonant vibrations in cladding that spans larger than typical distances should be examined. Considering the possible stress reversal caused by the highly fluctuating wind loads, fatigue-induced loosening of connections or even failure may occur. Attention to fatigue loading is especially important in hurricane regions.

Pressures are usually measured and reported in terms of pressure coefficients, defined as

$$C_{P_{\max}} = \frac{P_{\max}}{q_{\text{ref}}},\ C_{P_{\min}} = \frac{P_{\min}}{q_{\text{ref}}},\ C_{P_{\text{mean}}} = \frac{P_{\text{mean}}}{q_{\text{ref}}},\ C_{P_{\text{rms}}} = \frac{P_{\text{rms}}}{q_{\text{ref}}} \qquad \text{(C3-2)}$$

$P_{\max}$ = maximum instantaneous pressure measured over the sampling period;

$P_{\min}$ = minimum instantaneous pressure measured over the sampling period;

P_{mean} = temporal mean pressure;

P_{rms} = root-mean-square pressure $\left(\overline{(P - P_{\text{mean}})^2}\right)^{1/2}$;

$q_{\text{ref}} = 1/2\rho U_{\text{ref}}^2$ = dynamic velocity pressure associated with the mean wind speed, U_{ref}, at reference height;

an overbar indicates time average; and

all pressures are differential pressures with respect to the static pressure in undisturbed flow.

C3.3 AREA-AVERAGED PRESSURES

Averaging of pressures may be used to assess loads acting over larger areas or net loads acting across structural elements such as parapets or roof pavers. Several techniques for this have been devised. These efforts have been largely driven by the cost and physical size of early pressure instrumentation.

Area-averaged pressures are often measured for areas larger than 108 ft^2 (10 m^2) (or smaller for very high load gradients). Measurement techniques have been presented by a number of investigators, including Stathopoulos (1982), Surry and Stathopoulos (1977), and Gumley (1983). The averaging can be uniform, with all individual pressures considered equally, or weighted so that some parts of the panel or tributary area are given greater importance. For example, in the case of the wind load on the tributary area of a beam or purlin, the pressures on individual parts of this area can be weighted so that the spatial average gives a direct measure of the midspan bending moment. More direct measurement of area pressures has been achieved using porous polyethylene materials in the model construction (Roselas 1983).

C3.4 INTERNAL PRESSURES

The significance of internal pressures has often been neglected, and this has contributed to building failures. Wind-induced internal pressures depend on (1) the exterior pressure distribution; (2) the size and distribution of small, unavoidable cracks and leakage paths; (3) the size and distribution of large orifices such as doors or windows, which are open by tenant operation, or by wind-induced breakage; (4) the volume of enclosed air mass; and (5) the flexibility of the skin. For buildings controlled by (1) and (2), the internal pressure is determined by leakage (infiltration) and the air ventilation system; internal pressures defined by ASCE 7 are frequently used. The influence of large openings can have a significant effect on net skin pressure and on structural frame loads; these are frequently measured in the wind tunnel test.

Internal pressure can be experimentally determined, such as by Stathopoulos et al. (1979), and may also be used for the valuation of pressure differential across internal partitions. Wind can induce pressure within other partially sealed volumes, such as spaces under loosely laid roofing material and ballast pavers, and in cavities behind vented exterior wall panels. These internal or cavity pressures must be allowed for in situations where they can add to the wind pressures on surfaces behind the porous surface. For roof systems with impermeable membranes, and for rain screen-type curtainwalls, the venting of cavities behind exterior components can reduce the wind pressures acting on exterior surfaces by transferring a portion of the load to the underlying surface. Wind tunnel tests, including a representative simulation of the permeability of the exterior envelope, are required to ascertain the degree of pressure equalization. Because of the small geometry of these systems, larger scales or credibility checks with full-scale tests may be needed.

Chapter C4

OVERALL WIND LOADS (EXCLUDING AEROELASTIC EFFECTS)

There are several techniques for assessing overall wind loads on structures. The most appropriate technique for a particular case is determined by such factors as the target performance of the structure, the novelty or uniqueness of the structural design, the accuracy expected in the results, the relevance of the results to issues of public safety, and the time and funds available to complete the study. Where aeroelastic effects are not significant, direct measurement of wind loads on rigid models provides an economical and accurate methodology.

Electronic measurement and computing technology directly related to wind tunnel measurement techniques have advanced considerably and continue to do so. Consequently, approaches that were simply not viable previously are now being successfully developed, such as large-scale measurements and online integration of pressure, the recording and maintaining of large numbers of digital time series, and the combination of wind tunnel data with numerical computation. Because of this changing state of the technology, the following discussions focus more on the general physical and theoretical bases of different approaches and less on the particular mechanics of implementation.

C4.1 FORCE-BALANCE TECHNIQUE

C4.1.1 High-Frequency Force-Balance. The use of the high-frequency force-balance technique (often referred to as the base-balance technique) developed by Tschanz (1982) and, based on an earlier approach used by Whitbread (1975), is a widely accepted technique for wind tunnel model studies. The technique involves the use of a very stiff high-frequency balance model system which models only the exterior geometry of the structure. The wind tunnel study may be carried out at a stage in the design when only the exterior geometry of the structure has been fixed. When the structural properties become available, they are combined analytically with the wind tunnel data to determine full-scale responses. The measured quantity in the aeroelastic procedure includes the resonant response. In the force-balance technique, on the other hand, it is the spectrum of modal forces that is measured experimentally. The final response of the structure to that modal force is then determined analytically. Changes in the structural properties can be readily accommodated by iteration of the analytical procedures. Parametric studies, wherein the responses are predicted as functions of the structural parameters, are often feasible. Importantly, it is unnecessary to retest a new wind tunnel model unless significant changes in the exterior geometry are made.

C4.1.2 Principles and Assumptions. The fundamental premise of the force-balance technique is that the generalized or modal forces due to the wind can be estimated from the measured base moments experienced by a stationary model. The *modal force* is defined as the integral of the applied force weighted by the mode shape at the point of application.

A fortuitously similar quantity to the modal force occurs in the more easily measured base overturning moment. In this case, the applied forces incur a weighting naturally through the moment arm influence line, which varies linearly with height. When the mode shape is proportional to the influence function, the modal force and the base overturning moment are also proportional.

A similar approach can be taken for the twist modal force and modal torque. In this case, the loading is the torque per unit height, but the base torque influence line is unit constant over the height of the structure, which is less representative of twist mode shapes.

The base moments, including two overturning moments and the base torque, represent direct measurements of the modal forces on a structure when the following conditions are met:

1. The first three natural sway modes of the structure are decoupled and geometrically orthogonal in two sway directions and one twist direction.
2. The fundamental sway mode shapes are linear functions of height and pivot at a point where the moments are measured.
3. The fundamental twist mode shape is constant over the height of the structure.
4. There are no significant motion-induced forces involved and so the nature of the forcing remains the same on a responding structure as it is on a stationary structure.
5. The balance model system is essentially rigid, with a high natural frequency, so that the measured moments are not significantly amplified by the mechanical admittance of the system in the frequency range of interest.

In practice, these conditions are not usually fully met, but in most structures they are sufficiently satisfied to produce results useful for engineering design. Although only the first mode in each coordinate direction can be estimated, this is a minimal restriction because higher modes seldom contribute significantly. Load information over the height is derived from the mean pressure distribution obtained from the companion study of cladding pressures.

C4.1.3 Adjustments for Base-Balance Mechanical Admittance. It is necessary that the base moments used to represent the modal forces be moments as measured on a nearly rigid model. If the balance-model system responds dynamically to the wind loading, then the measured base moments will include the inertial loading effects of the system itself. If the motions are large, the aerodynamic interaction of the model with the wind could also contaminate the measurements. An attempt should therefore be made to make the balance-model system as rigid as possible while still being sensitive enough to provide reasonable signal strength. In this way, the frequency range of interest falls at the low end of the mechanical admittance function of the balance-model system where the dynamic amplifications are small. In some cases, the natural frequency of the system cannot be raised sufficiently high enough and the base moment measurements are amplified. There is, in principle, no difficulty in adjusting the spectral density measurements to account for this, provided the mechanical admittance of the system is well identified and may be treated as linear and uncoupled. It is always assumed, however, that the frequency is still high enough that the model motion is insignificant.

C4.1.4 Adjustments for Mode Shape. The assumption of a constant twist mode shape with height is never true for real structures. A typical twist mode shape for a very tall building lies somewhere between a constant and a linear function of height. In practical cases, the majority of the contribution to the torsional modal force comes from the upper half of the structure. The measurement of the base torque from the base balance may be made more representative of the generalized torque by artificially sheltering the lower portion of the model in such a way that the aerodynamic interference of the sheltering device is minimal. In many cases, a value of about 0.70 represents an adequate empirical adjustment factor to apply to measured base torques.

Although approximating the sway mode shapes by a linear function is sufficient in many cases, tall buildings and structures having mode shapes that significantly deviate from this approximation are not uncommon. The variation in mode shape can have a significant effect on the similarity of the modal force and the measured base moment spectral densities. Measurements aimed at quantifying these effects have been carried out by various researchers. The effect that typical deviations from a linear mode shape function has on final predicted responses differs depending on the type of response considered. Typical corrections for base moments might be ± 10% for bending and 30–40% (reduction) for torsion. Corrections to predicted accelerations near the top of a structure due to nonlinearly varying mode shapes may reach 20%. Further discussions of the effect of mode shape on building response estimates derived using the force-balance technique are given in Vickery et al. (1985), Holmes (1987), and Boggs and Peterka (1989).

C4.1.5 Treatment of Coupled Degrees of Freedom. Many structures, particularly buildings of more complex design, exhibit coupling between sway and twist motions. In most cases, coupling can be accommodated during the test by forming modal loads from coordinate-oriented measurements at each instant in time. For some cases, there may be a technical advantage in testing a multi-degree-of-freedom aeroelastic model. The trade-off is an improved measurement of modal loads and an estimate of aerodynamic damping in exchange for a significantly increased cost, increased test time, and a reduced measurement precision. For most buildings, the base moment balance (force-balance) technique provides sufficiently accurate measurements and is the method most used.

C4.1.6 Multilevel Force-Balance. An extension of the high-frequency force-balance technique is the multilevel force-balance [e.g., Reinhold and Vickery (1990)]. By measuring moments and shears at multiple levels, an improved estimate of the modal loads may be obtained, particularly in cases where the real mode shapes are not well represented by the idealized mode shapes inherent in the base-balance approach. This is particularly true of torsion about the vertical axis of the structure and nonlinearly varying sway modes.

The multilevel force-balance technique has the advantage that it does not require extensive pressure transducer equipment; however, it does require a specialized balance and more complicated models than its alternatives.

C4.1.7 Forced Oscillation. The forced oscillation technique is used to determine information about the motion-induced or aeroelastic forces acting on a structure (Steckley 1989). These are the wind forces on a structure that are induced by the motion of the structure through the air. In absolute terms, these forces are often relatively small compared with the random wind forces and the inertial and elastic forces of the structure itself. However, because they are correlated with the structural motion, they can be thought of as wind-induced damping and stiffness forces, which effectively modify the structural damping and stiffness.

C4.2 INSTANTANEOUS SPATIAL PRESSURE INTEGRATION

The ability to simultaneously measure large numbers of local pressures has prompted the development of techniques that use pressures measured over much larger areas of the structure or even the entire structure. The approach of integrating individually measured local pressures to assess modal loads is really an expansion of the online averaging approach described in the previous section. It offers an alternative to the force-balance approach of measuring modal loads, but is more widely applicable than that approach, which is suited only for tall, slender structures with relatively simple modal dynamics. The potential exists to apply the technique to a wide variety of structures, such as stadia and even bridges, and to handle the higher-frequency modes and complex coupled modes as well.

Proper measurement of overall loads requires a large number of simultaneously measured local pressures to be used to resolve the pressure field acting on the exterior surface of the entire structure. The advent of solid-state pressure-scanning instrumentation has made this a viable task; however, the actual extent of pressure tap coverage required remains an issue of debate. Opinions on the number of pressure channels for a typical high-rise building model, for example, range from 200 to as high as 1,000. A few points should be kept in mind in evaluating the required number of local pressure measurements:

- In evaluating overall mean loads, the spatial resolution required is greater because the static pressure field over the exterior surface of the structure must be adequately resolved and may have areas of high pressure gradients, particularly near corners and other flow separation points. On the other hand, mean loads do not require that the measured pressures be simultaneously occurring, so pressures from sequential tests may be combined.
- The dominant contribution of any modal load is due to pressure fluctuating at the modal frequency. Fluctuating pressures are generally correlated over distances inversely related to the frequency; in other words, the higher the fluctuating frequency, the more localized the fluctuating pressure. There are notable exceptions to this, such as vortex shedding wherein the fluctuations become highly correlated over a narrow range of frequency. Nevertheless, for many large building structures, the correlation distances of the pressure fluctuating at the important modal frequencies (i.e., the first two or three modes) are of the same order of magnitude as the structure itself. This fact helps to reduce the spatial resolution of measurements required for assessing modal loading.

Until further experience is accumulated, the requisite number of simultaneously measured pressures must be judged with particular care and on a case-by-case basis. The modal loads acquired from this technique may be accumulated and analyzed in either the time or frequency domain. The methodology is discussed in detail in Kareem (1982), Steckley et al. (1992), Irwin et al. (1995), Ho et al. (1999), and Isyumov et al. (2001).

Chapter C5
AEROELASTICALLY ACTIVE STRUCTURES

C5.1 GENERAL

A flexible, lightweight, lightly damped structure subjected to wind forces can undergo motions of sufficient magnitude to alter the wind forces acting on it. The effects of this feedback from structural deformation to the wind forces are called *aeroelastic effects*. Aeroelastic effects can either increase or decrease the structure's response to wind forces. In cases where they increase the response, they may result in aerodynamic instability, that is, large-amplitude, unstable oscillations. To simulate aeroelastic effects on a wind tunnel model requires not only that the geometry of the model be correct but also that the model's mechanical properties of flexibility, mass, and damping correctly simulate the corresponding full-scale mechanical properties. This type of model is called an *aeroelastic model*, and it is designed and constructed using the principles of dynamic similarity described in Section C5.1.2.

Examples of structures for which aeroelastic effects may be important are long-span bridges; tall, slender buildings; slender towers and chimneys; thin-walled shell structures; flexible, long-span roofs or canopies; and flexible, membrane-type structures. Structural components such as long, thin-walled truss members, slender column supports, and cables may also be subject to excessive vibrations caused by aeroelastic effects. Where analysis and previous experience leave doubt about the aerodynamic stability of the structure, or where it is desired to include the effects of aerodynamic damping in predictions of motion amplitudes and wind loading, aeroelastic tests should be undertaken. Aeroelastic tests can also be useful for structures with complex, three-dimensional (3D) modes of vibration that may present difficulties for rigid model methods.

C5.1.1 Types of Aeroelastic Instability. Because the primary motivation for aeroelastic tests is usually to ensure that a structure will not be subject to excessive dynamic response caused by aeroelastic instability, the more common mechanisms of instability will be briefly described. These are vortex excitation, galloping, and flutter.

Vortex Excitation. This form of instability originates from the regular shedding of vortices from both sides of a bluff structure. The shedding of vortices generates fluctuating forces on the structure, primarily perpendicular to the wind direction. The frequency of the vortex shedding, N_s, is proportional to the wind speed, U, and inversely proportional to the dimension L_b of the structure perpendicular to the wind.

$$N_s = S\frac{U}{L_b} \tag{C5-1}$$

where the constant of proportionality S = the Strouhal number. The Strouhal number depends on the cross section of the structure. For curved shapes, it may also depend on the Reynolds number. As the wind speed increases, so N_s increases according to Eq. (C5-1). When the wind speed reaches a value that causes N_s to match one of the structure's natural frequencies, resonant oscillations will build up. Once the vortex-induced oscillations are established, the shedding frequency tends to become locked to that of the structure's motion, even if the wind speed subsequently varies somewhat from the original resonant value. Therefore, vortex shedding oscillations typically are found to persist over a range of speeds close to the resonant value, reaching maximum amplitude at a particular speed. Depending on the magnitude of the maximum oscillations, vortex shedding can range from being of no consequence to causing serious problems, such as metal fatigue in structural components or unacceptable comfort conditions for users of the structure.

Galloping. Galloping is a dynamic instability of long, slender structures that can be traced to the way the across-wind mean force coefficient, C_y, varies with the wind angle, α. The across-wind force coefficient is defined as

$$C_y = \frac{Y}{\frac{1}{2}\rho U^2 b} \tag{C5-2}$$

where Y = the across-wind force per unit length in the positive y direction, ρ = air density, and b = the reference width of the structure in the direction of the wind. The wind angle, α, is defined to be positive when the y component of wind velocity is positive.

If the slope $dC_y/d\alpha$ of the across-wind force coefficient is positive, it can be shown that the aerodynamic damping of across-wind motions is positive, and therefore the structure will be aerodynamically stable against across-wind motions at all wind speeds. However, if $dC_y/d\alpha$ is negative, as happens on some cross sections (e.g., a square) over a limited range of α, then the aerodynamic damping becomes increasingly negative as the wind speed increases and, when its magnitude exceeds that of the natural damping of the structure, aerodynamic instability begins. Very-large-amplitude destructive oscillations can occur as a result of the galloping type of instability.

Flutter. Flutter is an instability that involves a combination of torsional and across-wind oscillations. A full description of this type of instability requires an examination of the equations of motion of across-wind and torsional motion, including the aerodynamic coupling terms; refer to Simiu and Scanlan (1996). Qualitatively, the flutter type of instability is somewhat similar to galloping in that negative aerodynamic damping is generated, which causes large-amplitude oscillations to build up once a certain critical wind speed is reached. However, flutter, in contrast to galloping, cannot be analyzed by use of derivatives of mean force coefficients, but must be analyzed by acquisition of the motion-induced flutter derivatives. Flutter is an instability that is of particular concern for long, slender bridge structures but has the potential to occur in other types of structures if they are flexible in torsion.

C5.1.2 Effects of Turbulence. As is evident from Chapter 2, the wind affecting most structures is turbulent at high wind speeds. The turbulence buffeting that occurs will cause motions of even aeroelastically stable structures. However, aeroelastic effects can alter the response to buffeting by changing the damping either positively or negatively, depending on the sign of the aerodynamic damping. Also, turbulence can have a significant influence on aerodynamic instabilities. For example, it has been found that turbulence tends to suppress vortex shedding excitation and to shift the onset of flutter instability to a higher wind speed. Therefore, in aeroelastic testing it is important to consider the full range of turbulence conditions that a structure could experience. For instance, vortex excitation of long-span bridges can occur at relatively low wind speeds. At these speeds over cool water, thermal stability effects can suppress turbulence in the atmospheric boundary layer, resulting in very low turbulence intensities. The wind tunnel tests must, therefore, consider the possibility of low turbulence intensities.

C5.2 AEROELASTIC MODELING PROCEDURE

C5.2.1 Requirements for Dynamic Similarity. The aeroelastic model must be capable of simulating the important responses to wind, which typically involves the mean deflections and the dynamic response in the lowest few modes of vibration. A full aeroelastic model simulates the mean deflections and all the important lower modes of vibration of the entire structure. A second approach is to use a partial model, which allows the response in one mode at a time, or in a reduced number of modes, to be examined in the wind tunnel. Analytical methods are then required to synthesize the complete response in all important modes. A third approach is a partial model used to extract the motion-induced flutter derivatives. Analytical methods are then used to determine the response. Modal similarity is not required in this case. A spring-mounted sectional model of a long-span bridge is an example of the second and third types of models.

To achieve dynamic similarity between a full aeroelastic model and full scale, it is necessary that certain nondimensional parameters have the same value on the model as at full scale. These may be deduced through dimensional analysis combined with a physical understanding of the structural system. The scaling parameters required to model the behavior of the natural wind are described in Chapter 2 and the associated commentary. The additional scaling parameter relationships pertaining to an aeroelastic model are as follows.

Mass Scaling. To achieve proper mass scaling, it is important for the model ratio of bulk structural density to air density to be the same as at full scale. This requires that the following relationships be satisfied:

$$\frac{M_m}{\rho_m L_m^3} = \frac{M_p}{\rho_p L_p^3}, \quad \frac{I_m}{\rho_m L_m^5} = \frac{I_p}{\rho_p L_p^5} \qquad \text{(C5-3)}$$

where M = mass or generalized mass, I = polar moment of inertia or generalized polar moment of inertia, ρ = air density, and L = reference dimension of the structure. The subscripts m and p denote, respectively, model and prototype (i.e., full-scale) quantities. For a sectional model, these relationships become

$$\frac{m_m}{\rho_m L_m^2} = \frac{m_p}{\rho_p L_p^2}, \quad \frac{i_m}{\rho_m L_m^4} = \frac{i_p}{\rho_p L_p^4} \qquad \text{(C5-4)}$$

where m and i are the mass and polar inertia per unit length. For roofs and membranes, the structural mass per unit area, m', becomes the relevant mass quantity. Mass similarity requires that

$$\frac{m'_m}{\rho_m L_m} = \frac{m'_p}{\rho_p L_p} \qquad \text{(C5-5)}$$

However, in cases where the structural mass is very small, such as for lightweight membrane structures, the added mass effect of the air adjacent to the surface becomes dominant. Where the added mass effect is dominant, exact matching of the structural mass is less critical.

Stiffness Scaling. For the simple example of a thin-walled tube where all the model dimensions, including wall thickness, are scaled down from full scale by the same ratio, the requirement that elastic forces be correct in relation to aerodynamic forces implies that the Cauchy numbers be matched, as in

$$\frac{E_m}{\rho_m U_m^2} = \frac{E_p}{\rho_p U_p^2} \qquad \text{(C5-6)}$$

where E = Young's modulus of elasticity, ρ = air density, and U = wind velocity. The subscripts m and p again denote, respectively, model and prototype (i.e., full-scale) quantities.

However, it is usually not necessary or practical to scale down all geometric dimensions, such as the wall thicknesses of structural components. Rather than matching the Cauchy number directly, related parameters based on overall structural rigidity are matched. There are various types of structural rigidity that it may be necessary to simulate on the aeroelastic model; for example, for cables, the rigidity EA for elongation is important, where E and A are the effective modulus of elasticity and effective cross-sectional area, respectively; for flexural members, the flexural rigidity EI is the relevant quantity, where I is the second moment of area; the torsional rigidity GK, where G is the shear modulus and K the torsional constant, may be important for some structures; for structures, the behavior of which is influenced by the warping rigidity, EC_w needs to be scaled where C_w is the warping constant. The following relationships must be satisfied for any of these rigidity parameters that directly affect the structural behavior:

$$\frac{(EA)_m}{\rho_m U_m^2 L_m^2} = \frac{(EA)_p}{\rho_p U_p^2 L_p^2}, \quad \frac{(EI)_m}{\rho_m U_m^2 L_m^4} = \frac{(EI)_p}{\rho_p U_p^2 L_p^4}, \quad \frac{(GK)_m}{\rho_m U_m^2 L_m^4} = \frac{(GK)_p}{\rho_p U_p^2 L_p^4},$$
$$\frac{(EC_w)_m}{\rho_m U_m^2 L_m^6} = \frac{(EC_w)_p}{\rho_p U_p^2 L_p^6} \qquad \text{(C5-7)}$$

Damping. Similarity of dissipative or damping forces in the structure requires that the damping ratio on the model be the same as at full scale, as in

$$\zeta_m = \zeta_p \qquad \text{(C5-8)}$$

where ζ = damping ratio. In most cases, the full-scale damping ratio is not accurately known; therefore, it is advisable to perform model tests for a range of damping values.

Gravitational Forces. For some structures, such as suspension bridges, the effect of gravity on their stiffness is significant. On tall towers, the P-delta effect is caused by gravity and, if it is desired to incorporate this secondary effect in the model simulation, then gravity forces need to be considered in the scaling of the model stiffness. To obtain the correct ratio of aerodynamic force to gravitational force, it is necessary to match the model Froude number to full scale, as in

$$\frac{U_m^2}{L_m g_m} = \frac{U_p^2}{L_p g_p} \qquad \text{(C5-9)}$$

where g = gravitational acceleration.

Velocity, Time, and Frequency Scaling. If Froude number scaling is used so as to accurately simulate gravitational effects, then—because the effective value of g is the same on the model as for full scale in all normal boundary-layer wind tunnels—the matching of Froude numbers implies matching of U^2/L. As a result, the velocity scale is given by

$$\frac{U_m}{U_p} = \sqrt{\frac{L_m}{L_p}} \quad \text{(C5-10)}$$

Once the velocity scaling has been determined by this relationship, then the scaling of the model stiffness becomes fixed through the rigidity parameters previously described.

In cases where gravitational forces are not important to the structural response, the velocity scale is no longer constrained by Froude number similarity and the velocity scaling can be set by the model modal stiffness rather than vice versa. The model stiffness and mass distributions result in it having certain natural frequencies of vibration. Denoting natural frequency by N, the nondimensional form of velocity may be written $U/(NL)$. The relationship between model and full-scale velocity follows from the matching of nondimensional velocities:

$$\frac{U_m}{N_m L_m} = \frac{U_p}{N_p L_p} \quad \text{(C5-11)}$$

Therefore, the ratio of model to full scale velocity is given by

$$\frac{U_m}{U_p} = \frac{N_m L_m}{N_p L_p} \quad \text{(C5-12)}$$

When Froude number similarity is not applicable, the model stiffness, and thus its natural frequencies, may be set to any convenient value, but the resulting range of test velocities must be compatible with the capabilities of the wind tunnel and its instrumentation, and must not result in excessively low test Reynolds numbers.

When Froude scaling is used, the model frequency scaling becomes fixed and is given by

$$\frac{N_m}{N_p} = \sqrt{\frac{L_p}{L_m}} \quad \text{(C5-13)}$$

When Froude scaling is not applicable, the model frequency scaling may, within the limits discussed previously in connection with velocity scaling, be selected at will. The time scaling is simply the inverse of the frequency scaling:

$$\frac{t_m}{t_p} = \frac{N_p}{N_m} \quad \text{(C5-14)}$$

where t = time.

Reynolds Number Similarity. As with other types of wind tunnel tests, it is not normally possible to satisfy the requirement of Reynolds number similarity. Reynolds number similarity ensures that viscous effects will be accurately modeled and it requires that

$$\frac{\rho_m U_m L_m}{\mu_m} = \frac{\rho_p U_p L_p}{\mu_p} \quad \text{(C5-15)}$$

where μ = viscosity of the air. For most structures, failure to satisfy this relationship is not of concern because they have sharp edges which dictate the airflow patterns. However, for circular cross sections and others with curved surfaces, the lack of Reynolds number similarity could lead to significant differences between model and full-scale response. Tests on Reynolds number-sensitive structures should be corrected for such differences based on available full-scale measurements, or based on relevant test data obtained at Reynolds numbers in the full-scale range. For structures with small components, such as railings or thin truss members, the Reynolds number of the individual components may fall to very low values (e.g., 100 or less) based on width, particularly when Froude scaling is used. In such cases, it is advantageous keep the component Reynolds number high by reducing the number of elements such as rails and increasing their width, keeping the exposed area the same.

C5.2.2 Types of Aeroelastic Models.

Tall Buildings and Slender Vertical Structures. For tests on a tall building, it would typically be necessary for the model to simulate at least the lowest three modes, that is, the lowest sway mode in each of the two principal directions and the lowest torsional mode. A lumped-mass type of model, with masses lumped at between three and seven heights and connected by appropriate flexures, would typically be needed to achieve this. The model shell is broken into segments connected to the lumped masses, thereby presenting the correct external shape to the wind without contributing to the stiffness. Examples of this and other types of models are described in ASCE MOP 67, *Wind Tunnel Tests on Buildings and Structures* (ASCE 1999), Templin and Cooper (1980), Isyumov (1982), Irwin (1982), and Zhou and Kareem (2003). If it has been established, through rigid model tests, for example, that the torsional response is not critical, then only the lowest sway mode in each principal direction would need to be simulated; in many cases, the sway mode deflection shapes are close enough to linear to be well represented by a rigid model rotating about two horizontal axes at its base. The elastic constraint in this last case is provided either by a spring system or by a flexure system at the model's base. The damping is also supplied by viscous or other types of dampers incorporated into the mounting system. Typical length scales for aeroelastic models of tall buildings are in the range of 1 : 200 to 1 : 600 and the model responses are measured using strain gauges, accelerometers, and displacement transducers.

Other vertical structures, such as towers, masts, and chimneys, can be modeled using similar techniques as for tall buildings. If they are cylindrical or if they involve fine lattice structure, then Reynolds number corrections will need to be considered. Froude number similarity may be important for some guyed structures.

Long-Span Bridges. For a long-span bridge it is important to establish the basic aerodynamic stability. Frequently this can be achieved through sectional model testing, which is described in the next paragraph. Also, by appropriate analysis methods combined with the sectional model data, the bridge's response to turbulence buffeting at high wind speeds may be estimated. In cases where the bridge is in hilly terrain or where other 3D effects are important, as may be the case during construction, a full aeroelastic model should be tested. Discussions of the different methods of testing of bridges, and the associated analysis methods, may be found in Scanlan (1992), Irwin (1992), Davenport et al. (1992), and Simiu and Scanlan (1996). For a very long–span bridge, a large wind tunnel is needed if a full model is to fit into the working section and the scale is not to be too small to maintain acceptable Reynolds number and geometric accuracy. For example, a wind tunnel more than 131 ft (40 m) wide has been used for this purpose (Miyata et al. 1992). An alternative approach which does not require such a wide wind tunnel is to construct a partial aeroelastic model, such as a taut strip model, rather than a model of the whole bridge; refer to Davenport et. al. (1992), for example. The taut-strip type of model consists of tensioned wires (or tubes if it is a taut tube model) upon which is mounted a segmented model of the deck structure. This model allows the response of the deck structure to a turbulent wind to be evaluated without the need to model the whole structure. Appropriate analysis methods are required to predict the behavior of the complete bridge from the measured response of the taut-strip model.

Because of their simplicity and the large scale to which they can be built, sectional models are commonly used for assessing the stability of a bridge. Sectional model scales are typically in the range of 1 : 10 to 1 : 100, depending on the size of the bridge and wind tunnel dimensions. The sectional model length should be approximately 4 times the model deck width. The model is mounted on a spring suspension system at its ends, and the damping of the model should be adjustable through a range representative of expected full scale values. The vertical and torsional responses are measured using strain gauge instrumentation in the spring system, or using other instrumentation such as displacement transducers or accelerometers. For the assessment of vortex excitation it is important to step through the wind velocity range in fine increments to ensure that the resonant velocities are properly identified. It is often important to establish how the vortex excitation amplitude varies with damping. Because vortex excitation is, in many cases, strongly affected by turbulence, it is also advantageous to study the effect of turbulence on the model response.

A complete simulation of turbulence is not possible on a sectional model, but an acceptable conservative approach for the purposes of examining vortex excitation is to provide a partial turbulence simulation. The partial turbulence simulation must match the full-scale nondimensional turbulence power spectra for both longitudinal and vertical velocity components for values of nondimensional frequency $nd/U = 1.0$ and greater, where n = frequency and d = the deck's vertical depth. The nondimensionalization of the power spectrum is $nS(n)/U^2$. For the assessment of flutter, two possible approaches may be used. The classic approach is to satisfy the mass similarity, damping similarity, and frequency similarity for the pair of modes being examined. The full-scale critical velocity is then directly calculated from the observed nondimensional critical velocity of the model. In cases where the critical torsional mode involves some accompanying lateral motion, the model mounting can be modified as described by Irwin (1979a) to allow this effect to be incorporated. The second approach is to use the method of Scanlan (Scanlan and Tomko 1971; Chen and Kareem 2004), which involves the measurement of six or more motion-related flutter derivatives on the sectional model, combined with an analysis in which the flutter equations are solved. Where the vertical mode shape differs significantly from the torsional mode shape, the Scanlan method will provide a more accurate determination of the flutter velocity. The flutter derivatives also prove to be of diagnostic value in assessing the inherent stability tendencies of proposed bridge deck shapes.

Flexible Roofs. The aeroelastic modeling of cantilever-type roof structures or canopies follows principles similar to those already described. Roofs over enclosed spaces involve several additional considerations. If the roof is over an airtight enclosure, then the pneumatic stiffness (from compression of the air) of the enclosed volume beneath it affects its response. The effect is to make the roof stiffer for any motions that change the enclosed volume beneath it (Tryggvason 1979a). However, in practice, the enclosure is typically not airtight but is ventilated by openings and fan systems, which result in pneumatic damping of any volume-displacing motions (Kind 1982). Also, volume-displacing modes will result in acoustic-type damping (Irwin and Wardlaw 1979). The additional scaling parameters involved are discussed in the cited papers. In tests on membrane-type structures, the model membrane can act like a microphone and respond to wind tunnel noise originating from fans, turning vanes, or other sources. Care may be needed to separate these spurious acoustic excitations from genuine aerodynamic excitation. For some cable-suspended roofs it may be necessary to use Froude number scaling (Kawakita et al. 1992).

The net result of pneumatic stiffness, pneumatic damping, and acoustic damping phenomena is to make it difficult for the wind to excite volume-displacing modes of vibration of roofs over enclosed spaces, but they can still cause significant motions in modes that do not displace volume. The natural frequencies of vibration of lightweight roofs can be significantly affected by added mass effects and damping effects (Irwin and Wardlaw 1979; Novak and Kassem 1990). Because the wind loading of large roofs is, in most cases, readily determined by multichannel pressure measurements, the usual purpose of an aeroelastic model is simply confirmation of aerodynamic stability. Aerodynamic stability may be established by measurements of the aerodynamic damping over the relevant wind speed range, with positive aerodynamic damping indicating a stable structure and negative aerodynamic damping indicating the presence of an instability.

Small Structures, Building Appendages, and Structural Members. Light standards, sign bridges, sunscreen structures attached to buildings, spires on top of buildings, antenna structures, slender truss members, and cables are all examples of small structures that can experience oscillations caused by aeroelastic effects. These oscillations can be identified and engineering solutions developed with the aid of aeroelastic models, designed using the scaling principles previously described, or, in some cases, by wind tunnel tests on the structure itself. For such structures, which are small enough to be completely enveloped by the turbulent gusts of the wind, it is in most cases not necessary or practical to model the complete spectrum of wind turbulence in the wind tunnel. Model scales may well be in the range of 1 : 1 to 1 : 20. Tests in smooth or partially simulated turbulent flows are typically sufficient, provided that the wind tunnel speed is regarded as a gust speed and that the range of wind angles tested covers the same range that will be experienced due to the vertical and lateral components of wind turbulence.

Chapter C6
EXTREME WIND CLIMATE

C6.1 GENERAL

To obtain the predicted loads or responses as a function of the return period, the results from the wind tunnel tests are combined with information on the extreme wind climate at the location of interest.

C6.1.1 Wind Types and Their Effects on the Wind Database. Throughout the various regions of the United States, extreme winds are governed by different meteorological phenomena. Along the Gulf and Atlantic coasts of the United States, extreme winds are driven by the influence of tropical cyclones or hurricanes. In the Great Plains and the southeastern portion of the country, thunderstorms dominate the extreme wind climate, and in the western mountain states the mountain ranges can induce large-amplitude atmospheric waves that may dominate the extreme winds. Over much of the remainder of the country, the extreme wind climate is dominated by the passing of large-scale extratropical storm systems. At any one location, the record of wind speeds measured by an airport anemometer is likely to contain winds produced by more than one of the above storm types. The present state-of-the-art prediction of structural loads and responses separates hurricane winds from other wind types, but thunderstorms and downslope winds are not, in general, treated any differently than the winds produced by the large-scale extratropical storms. Over most of the United States, the extreme winds are produced by hurricanes, thunderstorms, or large-scale, extensive pressure systems. A review of the aforementioned wind types is given in Golden and Snow (1991) and is briefly discussed in the following sections.

C6.1.2 Extratropical Storm Systems. Extensive pressure systems (EPS) or extratropical storm winds are produced by the large-scale high- and low-pressure systems that move across North America from the west. The strongest winds associated with these storms are produced by severe low-pressure systems that occur most often during the winter months. The winds in these extratropical storms are produced through a balance in the pressure gradient forces and the Coriolis force. The wind speed at the surface is reduced due to frictional effects, and the wind direction is changed, also a result of surface friction. The characteristics of the boundary layer associated with the extratropical storm winds are reasonably well understood, and it is these winds that are well modeled in the boundary-layer wind tunnel.

C6.1.3 Thunderstorms. Thunderstorms constitute a violent form of atmospheric convection and are generally composed of short-lived cells, consisting of regions of strong vertical air motion. With favorable conditions, a supercell may develop (Browning 1964), in which updrafts and downdrafts coexist for periods of an hour or more. The cold downdraft air often produces a sharp gust-front. Other strong winds produced by thunderstorms include those associated with downbursts and tornadoes. Thunderstorms tend to dominate the extreme wind climates in the middle portion of the United States (Twisdale and Vickery 1992, 1993).

C6.1.4 Hurricanes. In economic terms, the most devastating windstorms in the United States are those produced by hurricanes (tropical cyclones or typhoons, as they are referred to elsewhere). Although they are small compared with extensive pressure systems, and are relatively infrequent, when they make landfall in urban areas the damage they produce is unequaled by any other single wind event. Tropical cyclones occur most frequently in the late summer and early fall, when ocean temperatures and air humidity are the highest. Tropical cyclones originate over tropical ocean water near latitudes of 15°. Fully developed hurricanes have typical diameters in the range of a few hundred kilometers; however, the strongest damaging winds are concentrated near a few tens of kilometers from the center of rotation. In general, as a tropical cyclone becomes stronger, the size (or eye diameter) tends to decrease. In the United States, the most severe storms occur from late August to early October. Once the tropical cyclones pass over land, their intensity diminishes rapidly, as their source of energy from the warm waters is removed. As a result, for return periods of 100 years or less, tropical cyclones or hurricanes are generally only important for locations within about 62 mi (100 km) of the coast.

C6.2 ANALYSIS OF WIND DATA

C6.2.1 Background. Historical records of surface wind speed and direction in the United States are available in the form of 2 min averages recorded once per hour (or once per three hours) and daily maximum peak gust or fastest-mile data. The peak gust wind speeds have durations on the order of 2 to 3 s. The averaging time, T, in seconds, associated with the fastest-mile wind speed is

$$T = 3600/U_{fm} \qquad \text{(C6-1)}$$

where U_{fm} = the fastest-mile wind speed in mi/h. For fastest-mile wind speeds of engineering interest, the averaging time is in the range of 20–120 s. Two basic methodologies are used to combine the wind speed data with wind tunnel data to derive estimates of response as a function of return period. The first of these methodologies employs the hourly measurements of wind speeds and direction, combined with an upcrossing approach; the second method makes use of the annual extreme values directly. The methodologies are discussed in the two following subsections. Because the experimental data derived from wind tunnel tests are usually referenced to mean wind speed at gradient height, when using either the extreme value approach or the upcrossing approach, the surface wind speeds must be converted to equivalent hourly mean values at gradient height (or other reference height). This conversion is usually made assuming a power law relating the mean wind speed at surface level to the mean wind speed at gradient (reference) height.

In cases where the wind climate is to be determined using annual extreme values of peak-gust or fastest-mile data, these wind speed measurements must first be converted to equivalent mean hourly values. This conversion is usually made with the assumption that the wind speeds were produced by an extratropical storm, and that the relationships such as those developed by Durst (1960), relating the wind speed averaged over an arbitrary length of time to the wind speed averaged over one hour, are valid.

C6.2.2 Hourly Wind Speed Data: The Parent Distribution Approach. The use of hourly measurements of wind speed to define the parent probability distribution of wind speed and direction, coupled with an upcrossing approach to derive the extreme, was first discussed in Davenport (1968) and further examined by Gomes and Vickery (1977). Details of the upcrossing approach are presented in Davenport (1977, 1982, 1983) and Lepage and Irwin (1985). The upcrossing method can also be used with the probability distribution of wind speed and direction derived from upper-level (balloon) measurements of wind speed. Details of the methodology are described in detail in Vickery (1973). Balloon or upper-level wind speed measurements are available for a number of locations in the United States and, depending on the station, balloons are released either twice daily or 4 times per day. Measurements of wind speed and direction (1 min averages), temperature, and so forth are recorded at standard pressure levels (e.g., 950 mbar, 900 mbar). The main advantage of using the upper-level wind speed measurements is that the data are free from the influence of terrain effects, nearby vegetation, and structures. The main disadvantage of the approach is that the wind speed measurements taken only 2 or 4 times a day provide less data than those taken at surface level; consequently, there is more uncertainty built into the statistical models used to fit the data. Also, balloons may not be launched during high wind events.

C6.2.3 Wind Speed and Response Prediction Using Extreme Value Analysis. In instances where the annual extreme wind speed data are extracted from a continuous record of wind speed, the annual probability of exceeding a wind speed can be determined by fitting the wind speed data to a Type I extreme value distribution. The use of a Type I extreme value analysis of wind speeds at 129 stations in the United States given in Simiu et al. (1979) forms that basis for the nonhurricane design wind speeds given in ASCE 7-88. The Type I distribution is also used to determine the design peak gust wind speeds in the United States, as given in ASCE 7-95 (ASCE 1996), 7-98 (ASCE 2000), and 7-02 (ASCE 2003), which is based on the work of Peterka (1992) and Peterka and Shahid (1998). An alternate to the Type I fit has been presented by Simiu and Heckert (1996), which has not yet been incorporated into most analysis procedures. In the United States, the annual maximum wind speeds have been recorded in the form of either peak-gust or fastest-mile values. However, fastest-mile speeds are no longer recorded. For use with wind tunnel test data, the peak-gust or fastest-mile wind speed data must be converted to equivalent mean hourly values using the methodology described in Durst (1960) or the modified Durst method given in Simiu and Scanlan (1996).

In instances where only relatively short records of extreme winds are available, monthly extremes can be used to estimate the 50- or 100-year return period wind speeds (Simiu et al. 1982).

C6.2.4 Comparison of Wind Climate Models. As noted in ASCE MOP 67, *Wind Tunnel Studies of Buildings and Structures* (ASCE 1999), different wind data sources can lead to different estimates of the wind-induced response estimates. Because these differences can exist, it is often advisable to produce estimates of response versus return period using more than one representation of the local wind climate and present the largest values as design values.

C6.3 HURRICANE WINDS

Along most of the Gulf and Atlantic coasts of the United States, the design wind speeds are dominated by the influence of tropical cyclones or hurricanes. Reliable estimates of wind speed as a function of return period in these hurricane-prone regions cannot be derived using the more traditional techniques outlined in Sections 6.2 and C6.2. The traditional approaches cannot be used because at any given location there are very few direct measurements of hurricane wind speeds. These direct wind speed measurements do not exist because hurricanes are relatively small storms, they occur infrequently (at any one location), and in instances where direct measurements of wind speeds are potentially available, the maximum wind speeds produced by the storm are often not recorded because the anemometer fails or power is lost.

To overcome the difficulties associated with the lack of direct hurricane wind speed measurements at a prescribed location, an indirect method first proposed by Russell (1971) is used to determine design wind speeds produced by hurricanes. This method owes its success to the fact that, although hurricane wind speed measurements at a given location are not available, information on key parameters describing a hurricane, including central pressure, position, time data, and storm size, is available. Using these data, statistical distributions for central pressure, storm heading, translation speed, distance from the site, and radius to maximum winds can be derived. This information is then used in combination with mathematical models of a hurricane windfield to simulate many thousands of hurricanes near the site. The final result of the procedure is a large number of simulated wind speed traces generated from site-specific statistics of key hurricane parameters, thus retaining the local hurricane climatology.

Details of the use of mathematical simulation methodology as applicable to the United States are described in studies by Russell and Schueller (1974), Tryggvason et al. (1976), Batts et al. (1980), Georgiou et al. (1983), Georgiou (1985), and Vickery and Twisdale (1995a, b). Tropical cyclone wind speeds along the northern Australian coast have been studied by Gomes and Vickery (1976) and Tryggvason (1979b). Although the simulation methodologies used by these investigators are similar, there are significant differences in the statistical models, methods of data analysis, data sources, and hurricane windfield modeling used in each of these studies; consequently, wind speed predictions at a given location may vary from study to study.

C6.4 OTHER WINDS

Downslope winds at design level occur primarily as a consequence of wind blowing across mountains. Under critical circumstances, wind speeds on the lee side of the mountains can reach high levels at frequent intervals. Downslope winds tend to be exceptionally high only for a limited range of directions. Many, but not all, downslope wind regions are marked on the wind map of ASCE 7 as special wind regions. Some local wind data may have been obtained in some locations, but most do not have any wind records characteristic of the local area. Reference to local building officials or residents may be the only method to establish a design speed.

A good discussion of tornadoes and microbursts is contained in the commentaries to ASCE 7-95, 7-98, and 7-02. About half of all tornado speeds are less than about 100 mi/h (161 km/h). The design for severe tornadoes and microbursts is normally at recurrence intervals of 1,000 years or more, typically beyond the level of design attention. However, current atmospheric science literature continually increases the level of understanding of these phenomena, and their inclusion in a wind tunnel study is possible if the implications of the modeling (e.g., lack of understanding of the windfield structure) and long return periods are discussed with the client.

C6.5 COMBINATION OF WIND AND WIND TUNNEL DATA

Methods to combine wind data with wind tunnel coefficients include the upcrossing method (Davenport 1977, 1982, 1983; Lepage and Irwin 1985) and extreme value methods (Cook 1985; Holmes 1986).

In the time series or storm passages method, wind tunnel data are combined with the detailed hourly meteorological records of speed and direction to generate complete time series on an hour-by-hour basis of the building wind load or load effect (Isyumov et al. 2003; Gamble et al. 2001). Extreme value analysis is undertaken on the annual, monthly, or individual storm extremes of load or load effect to determine design conditions at an appropriate mean recurrence interval. A discussion of various methods is contained in Chapter 8 of Simiu and Scanlan (1996). Alternate methods may be acceptable but should be explained in the report.

Chapter C7
SNOW LOAD MODEL STUDIES

C7.1 INTRODUCTION

The snow loads on a building are strongly affected by the wind, which redistributes the snow particles through drifting. The highest snow loads on buildings are typically caused by drifting (Templin and Shriever 1982). The ASCE 7 snow load provisions account for the more common forms of snow drift, but scale-model studies are recommended to aid interpretation of the provisions and help define snow loads for cases where the roof shape differs from the shapes covered in the standard. Also, where surrounding structures, such as nearby taller buildings or topographical features, generate unusual wind flows over the roof, scale-model studies can be helpful in detecting unusual snow load distributions not identified in ASCE 7.

Model studies for snow loads have not been subject to as much full-scale verification as for wind loads, and there is a greater diversity of methodology. However, two basic approaches have been used to study snow drifting on scale models. In the first, the model is subjected to a simulated wind flow, either in a wind tunnel or a water flume, and particles are introduced into the flow to simulate the snow particles. A variety of materials have been used as the model particles, including sand, sawdust, wheat bran, sodium bicarbonate, glass spheres, activated clay particles, and ground walnut shells. The objective in this approach is to simulate the particle trajectories with sufficient precision so that realistic drift formations will appear on the model from which full-scale predictions may be made. Methods using this approach will be called Particle methods in this commentary. Examples of applications of Particle methods are described in Wuebben (1978), Iversen (1982), Irwin and Williams (1983), Anno (1984), Kind (1986), Petersen and Cermak (1988), Isyumov and Mikitiuk (1992), O'Rourke and Weitman (1992), and Irwin and Gamble (1992). The Particle method is best used for single storm events or where cumulative effects of changing wind directions and speeds, melting, rain on snow, and so forth are not important. Therefore, the Particle method is usually employed to provide qualitative or semiquantitative information to supplement other data from codes, standards, and field observations, rather than to provide independent quantitative predictions. Under limited conditions, such as where consolidation and melting are not significant, this technique has been used to simulate multiple-storm effects. Despite the Particle method's limitations in a quantitative sense, it can be useful in cases of complex geometry in alerting the design team to the presence of unusual snow drifts not provided for in codes and standards, and in helping to resolve uncertainties in interpreting code provisions.

A significant limitation of the Particle method is its inability to track the history of snow accumulation on the roof over weeks or an entire snow season. To overcome this limitation, a second approach has been developed that includes the effects of snow deposition at different directions on different days, and includes a heat and mass balance over time using meteorological records to determine snowfall rates, wind speeds, temperatures, and cloudiness that drive the heat and mass balance. This second approach is called the Time Step simulation method, and it has two forms:

1. The first, "Time Step A," does not attempt to simulate the particle trajectories on the model, but instead uses the model to determine the wind velocity patterns on the roof. The local wind velocities are measured at a large number of points near the model's roof for the full range of wind directions (typically at intervals of 10° or 22.5°) in a boundary-layer wind tunnel, and then computational methods are used to predict how the snow will drift in the velocity field. Examples have been described by Irwin and Gamble (1988, 1992), Gamble et al. (1991), and Williams et al. (1992). The computations proceed in a time step manner, using the detailed meteorological records of wind, snowfall, rain, temperature, and cloud cover. A mass and heat balance within each area element, including the transfer of mass due to drifting, is undertaken at each time step. In this manner, the cumulative effects of successive snowfalls, changing winds, rain, melting, and refreezing are accounted for. The framework of this method is such that the predictions account for many more factors than is possible with the Particle method alone. The output is in the form of time histories, on an hourly basis, of snow load patterns on the roof. Typically, the simulations are run for as many winters as there are detailed meteorological records, usually 30 to 50. This method may be supplemented by tests using the Particle method, with the results providing the overall mass in the drifts and the Particle method being used to help define the details of drift shape.
2. The second, "Time Step B," begins with use of the Particle method to define deposition amounts at a large number of locations for each wind direction where meteorological records indicate significant snowfall in the presence of wind. The snowfall amounts are normalized by the snowfall in an open country environment so that each test provides a nondimensional snowfall record for use in the numerical procedures to follow. The Particle tests are frequently run at more than one wind speed so that the effects of drifting at different wind speeds can be accounted for and the impact of drifting due to high winds immediately following a snowfall can be taken into account. The method has been described in Peterka and Esterday (2004). The computations proceed in a time step manner, similar to the Time Step A method, using the detailed meteorological records of wind, snowfall, rain, temperature, and cloud cover. The meteorological data are used with the nondimensioinal snowfall data from the Particle tests to allocate snowfall for each hour of a storm. A mass and heat balance at each measurement position is undertaken at each time step. In this manner, the cumulative effects of successive snowfalls, changing winds, rain, melting, and refreezing

are accounted for. The framework of this method is such that the predictions account for many more factors than is possible with the Particle method alone. The output is in the form of time histories, on an hourly basis, of snow load patterns on the roof. Typically, the simulations are run for as many winters as there are detailed meteorological records, usually 30 to 50. This method has a much shorter history than the Time Step A method.

Advantages and disadvantages of the three methods (Particle method, Time Step A, and Time Step B) are briefly listed. The Particle method is best for single storms or a few closely spaced storms where the effects of melting between storms are not important. Many locations in the United States have their 50-year events as single storms with little or no accumulation prior to the event; the Particle method works well for these cases. The two Time Step simulation methods have their advantages where snow accumulations contributing to the design magnitude occur over a number of storms separated by sufficient time that heat transfer effects are important. The Time Step A method can be performed without any particle deposition studies at all, but then relies on wind transport of snow that is generic and may not be based specifically on the geometry under study; this technique can be augmented with particle studies to assist in definition of deposition shapes, but inclusion of this data may not always be straightforward. The Time Step B method has the advantage of using particle deposition to define deposition distribution for each storm, but the disadvantage of not permitting drifting to occur after the initial storm; in many locations, snow consolidation will effectively prevent drifting after a short time, but in other locations, drifting can continue to occur for some time.

C7.2 MECHANISMS OF SNOW DRIFTING

Drifting is a complex process that has been studied by a number of authors, including Bagnold (1941), Dyunin (1963), Mellor (1965), Isyumov (1971), Kobayashi (1973), Tabler (1980), Kind (1981), Iversen (1982), and Schmidt (1986). At the most frequently occurring wind speeds [i.e., less than about 31 mi/h (14 m/s) near the snow surface], the bulk of the mass flux of snow particles occurs within a few centimeters of the surface through a process called *saltation*. Saltation consists of particles being ejected from the surface by the impact of landing particles. Once thrown up into the air stream, they are rapidly accelerated by the wind before they land and throw up yet more particles. Although the smaller particles do fly higher and can cause visibility problems, their mass flux is small and the trajectories of the heavy particles that contribute overwhelmingly to the mass flux are not more than a few centimeters high and a few tens of centimeters long. Saltation drifting begins once the wind speed exceeds a certain value. This threshold speed, U_{th}, is typically about 9 mi/h (4 m/s) for fresh snow [as measured about 3 ft (1 m) above the surface]. Old snow, especially after a melting episode or if it has been rained upon, is much less able to drift and has a much higher threshold velocity.

The mass flux of saltating particles is ultimately limited by the drag effect of the particles on the wind near the snow surface. The wind is slowed down enough that it can no longer sustain a higher load of airborne particles, and an equilibrium mass flux is reached. As discussed by Gamble et al. (1991), the distance required to reach the equilibrium flux is probably on the order of 16 ft (5 m) or less. The equilibrium mass flux rate, q, can be expressed as

$$q = f(U, U_{th}) \qquad \text{(C7-1)}$$

where U = the mean wind velocity at a reference height, usually on the order of 3 ft (1 m), and U_{th} = the threshold value of U for the onset of drifting. An example of this relationship is that of Dyunin (1963):

$$q = cU^2(U - U_{th}) \qquad \text{(C7-2)}$$

where the wind speed is in m/s, U_{th} is taken as = 4 m/s, and the constant $c = 3.34 \times 10^{-5}$ kg/(s²m⁴). Another relationship, suitable for fresh snow, is that given by Kobayashi (1973):

$$q = cU^3 \qquad \text{(C7-3)}$$

where $c = 3 \times 10^{-5}$ kg/(s^2m^4).

These relationships essentially describe saltation-type drifting. For example, the Kobayashi data were obtained by capturing the saltating snow in trenches. Smaller snow particles in suspension would not have been captured. The additional mass flux in suspension is very small at low wind velocities but becomes more significant at higher speeds. Suspension flux eventually becomes dominant when the root-mean-square of vertical velocity fluctuations in the turbulent wind exceeds the terminal velocity of the heavier snow particles; the indications are that typically this occurs when the mean wind speed is in the range of 50–65 ft/s (15–20 m/s) at a height of 3 ft (1 m) above the snow surface. Schmidt (1986) provides information on the division between saltation drifting and suspension drifting, which can be used to assess the probable contribution of suspension drifting to total snow flux.

An important implication of saltation drifting is that snow will not generally drift in any significant quantities from ground level up onto a roof because the particles drifting along the ground do not fly high enough. Therefore, because snow can usually drift off the roof but not up onto it, there is a general tendency for the overall amount of snow on the roof to be depleted by wind action. Exceptions to this may occur in extremely high winds, such as might be seen in an arctic blizzard or where the ground shape forms a "ramp" up to the roof level.

The wind velocity patterns on a roof depend on the roof's geometry and can be very complex. Snow will tend to accumulate in areas of decelerating flow and to be scoured away in areas of accelerating flow. In the wind tunnel or water flume, the full-scale wind velocity patterns can be reproduced at model scale provided an appropriate simulation of the planetary boundary layer (PBL) is obtained in the approaching flow. Chapter 2 of this standard describes the requirements for simulating the PBL. Section C7.3 deals with the requirements for simulating the drifting of snow particles in a given wind velocity field.

C7.3 SCALING REQUIREMENTS FOR PARTICLE METHODS

A number of researchers have studied the question of how to simulate the transport of snow using scale models. Examples are Odar (1965), Isyumov (1971), Iversen (1982), Irwin and Williams (1983), Anno (1984), Kind (1986), Peterka and Petersen (1990), Kwok et al. (1991), Isyumov and Mikitiuk (1992), O'Rourke and Weitman (1992), and Toyoda and Tomabechi (1992). Two important aspects must be considered. One is the simulation of the trajectories of falling particles in a wind flow. The other concerns the complex interaction of the air flow and the snow particles at the snow surface, which governs saltation drifting. Considering the simulation of falling snow particles first, it can be shown from the equations of motion of a particle in a fluid (Etkin 1971; Irwin and Williams 1983) that the following two parameters must have the same values on the model as at full scale:

$$\frac{w_t}{U}, \frac{U^2}{Lg\beta} \quad \text{(C7-4)}$$

where w_t = terminal velocity of falling particle, U = mean wind velocity at a reference point, L = representative building dimension, g = gravitational acceleration, $\beta = (\gamma - 1)/(\gamma + \gamma_a)$, γ = ratio of particle density to air or fluid density, and γ_a = additional contribution to γ due to the added mass effect. For a model in a wind tunnel, γ_a is essentially zero because in air the added mass effect is negligible, and β is effectively 1.0 because of the small density of air. For a water flume, γ_a is approximately 0.5, and if sand particles, for example, are used, β is also close to 0.5. The first of these parameters implies that the ratio of terminal velocity of the particles to fluid flow velocity must be the same on the model as at full scale. The second parameter is the densimetric Froude number, and matching of this parameter ensures that the ratio of aerodynamic force to gravitational force is correct on the model.

These two parameters are essentially all that need to be matched to full scale for problems where the interaction of the snow particles with the surface is not important. Examples of problems falling into this category are the initial deposition of falling snow in wind, tracking the snow particles falling off a high surface onto a lower surface, and some snow infiltration problems at air intakes.

However, most snow loading problems involve saltation drifting; this requires that the interaction of the snow particles with the snow surface also be considered. An important additional parameter governing the surface interaction is the ratio of mean flow velocity, U, to the threshold velocity, U_{th}, for the onset of drifting:

$$\frac{U}{U_{th}} \quad \text{(C7-5)}$$

This ratio should have the same value at model and full scale. An additional related requirement is that the mass flux relationship for the model should follow a similar functional form to the full-scale relationship. The flux relationship, Eq. (C7-1), may be expressed in nondimensional form:

$$\frac{q}{\rho_p h_s U_{th}} = cf\left(\frac{U}{U_{th}}\right) \quad \text{(C7-6)}$$

where ρ_p = particle density, h_s = height at which U is defined in the saltation flux relationship, and c = a dimensionless constant. h_s is typically 0.5 m to 1 m at full scale, which is much smaller than the typical building dimensions. On the model, h_s should also be much smaller than the model building dimensions. This will allow consistency with the full-scale observation that the saltation process is confined to a thin layer near the snow surface, and that the saltation trajectories are short in relation to the building dimensions. The function f should be identical for the model and full scale, but c can be allowed to differ.

Peterka and Petersen (1990) have shown that, for many geometries, the accumulation is relatively insensitive to Froude number, and thus to the ratio U/U_{th}. Therefore, the form of Eq. (C7-6) may vary in many cases.

The mass flux relationship may be used to establish an appropriate nondimensional time, $\bar{t}$ (Iversen 1982; Irwin and Williams 1983). In the latter reference it is expressed in the form

$$\bar{t} = \frac{cU_{th}h_s}{L^2}t \quad \text{(C7-7)}$$

where t = time. This nondimensional time has the same value on the model as at full scale, from which the ratio of prototype (full scale) time, t_p, to model time, t_m, is found to be

$$\frac{t_p}{t_m} = \frac{(cU_{th}h_s)_m}{(cU_{th}h_s)_p}\left(\frac{L_p}{L_m}\right)^2 \quad \text{(C7-8)}$$

where subscripts m and p denote model and prototype quantities, respectively. When the model test is run for a certain nondimensional time, the resulting drift depth, d, when expressed as a nondimensional depth, d/L, will be the same as at full scale if the other scaling requirements described above have been met. According to Eq. (C7-8), the model test duration should vary as the model size squared, that is, as L_m^2.

Kind (1986) suggests that strictly one or more of the following conditions should also be satisfied:

$$\frac{u_{*th}^3}{2g\nu} \geq 30 \quad \text{(C7-9)}$$

$$\gamma \geq 600 \quad \text{(C7-10)}$$

where u_{*th} = surface shear stress velocity at threshold of saltation and ν = kinematic viscosity. Inequality Eq. (C7-9) amounts to a minimum Reynolds number requirement, and inequality Eq. (C7-10) is a minimum density ratio that Kind found necessary in wind tunnel tests. In water flume tests, Eq. (C7-10) is not satisfied but, as is next discussed, realistic drift patterns can nonetheless be generated in many situations.

C7.4 APPLICATION OF PARTICLE METHODS

In practice, the requirement to match the parameters governing the trajectories of falling particles [Eq. (C7-4)] can be met without too much difficulty. For example, on a 1 : 200 scale wind tunnel model, to match the required densimetric Froude number would involve wind tunnel velocities equal to $1/\sqrt{200}$ of the full-scale values, and particles with a terminal velocity equal to $1/\sqrt{200}$ of the full-scale value. The velocities of the test are very low and the particles must have a very low terminal velocity. Nonetheless, such velocities can be readily achieved and suitable particles found. Similarly, in water flumes the dynamic similarity parameters in Eq. (C7-4) can be matched to full scale by appropriate choice of particles.

However, in wind tunnels, if the parameters in Eq. (C7-4) are satisfied, it then proves impossible to match simultaneously the parameters Eqs. (C7-5), (C7-9), and (C7-10) that affect saltation drifting. The velocity of the test becomes too low to obtain a match to the full-scale value of parameter Eq. (C7-5), or to initiate saltation drifting, which is typically the most important mechanism of drift formation. Therefore, in wind tunnel Particle methods, the decision is frequently made to relax the requirement for densimetric Froude number scaling and instead to concentrate on matching parameter Eq. (C7-5) while also satisfying Eqs. (C7-9) and (C7-10). Because of not matching the densimetric Froude number, the lengths and heights of the saltating particle trajectories become exaggerated compared with the model dimensions. The consequences of this can vary depending on the application. Good agreement has been demonstrated with full-scale observations of drifts around snow fences, for example, as demonstrated by Iversen (1982) and Kind and Murray (1982). However, the exaggerated particle paths cause difficulty in using the wind tunnel to duplicate some other types of full-scale drifts. Examples include the triangular drifts in the region of aerodynamic shade downwind of a roof step, or the drift on the downwind half of an arched roof aligned with its axis normal to the wind.

In water flume tests, the densimetric Froude number in Eq. (C7-4) and parameter Eq. (C7-5) are easier to satisfy simultaneously, and so the problem of exaggerated particle trajectories is avoided. Good simulations of the triangular drift downwind of a roof step can be obtained. Irwin and Williams (1983) describe a number of cases where water flume tests provided good qualitative predictions of snow drifts on and around buildings. However, because of the much higher density of water compared with that of air, the density ratio requirement Eq. (C7-10) must be relaxed, and this leads to some differences in the details of the saltation process. Thus, for example, when sand is used to simulate snow in the water flume, unrepresentative ripples or small dunes may form in smooth, open areas of the model.

From the preceding discussion of Particle methods, it is clear that they cannot be regarded as precise dynamic simulations in either the wind tunnel or the water flume, but rather as approximate analogues of the real drifting processes. Time scaling relationships can be derived, such as Eq. (C7-8), but the lack of exact dynamic similarity makes precision in time scaling difficult. Therefore, an acceptable and simple alternative to applying detailed time scaling relationships is to use the reference snow depth method. With this method, a location is selected as a reference point where the full-scale snow depth is assumed to be known (from code provisions, for example, or from direct full-scale observation) and where there is confidence that the model is producing a realistic simulation. The model test is run until the model nondimensional snow depth, d/L, at the reference location matches the known full-scale value. For example, in a water flume the reference location could be at a downwind-facing roof step, where ASCE 7 gives good guidance as to the expected 50-year peak snow load and snow depth in the triangular drift. A special reference block with a roof step set adjacent to the study model may be used to provide a well defined reference condition. When the model reaches the required depth at the reference point, the test is halted and the snow depths elsewhere on the model are assumed to be indicative of the full-scale drift depths that would occur under design conditions.

In Particle methods, the measurements involved are usually of the model snow depths. Depending on the accuracy needed, the equipment used for depth measurements can range from simple graduated rods to laser distance transducers [e.g., Toyoda and Tomabechi (1992)] or a Moiré fringe camera [e.g., Kwok et al. (1991)].

C7.5 APPLICATION OF THE TIME STEP SIMULATION METHOD

The Time Step A simulation method is less demanding in terms of model scaling than are Particle methods, because only the wind velocity field needs to be simulated. The methods for achieving this are well established and are described in Chapter 2 of this standard and the associated commentary.

The main task in the model test under the Time Step A method is to measure the mean wind speed and direction at many locations near the roof surface. Typically, a reference height corresponding to about 3 ft (1 m) above the full-scale roof would be selected for these measurements, and they would be carried out on approximately a 30 ft (10 m) (full-scale) grid pattern for at least 16 wind directions. Grid patterns more dense than 30 ft (10 m) may be used, but ultimately the grid density is limited by the equilibrium length for the saltation process, which, as explained earlier, is probably on the order of 16 ft (5 m). Special probes have been developed for these types of velocity measurement; refer to Gamble et al. (1991).

Once the velocity field in the Time Step A method has been mapped for the different wind directions, the prediction method proceeds from there onward by computation. The computer programming involved is complex; the details have been described by Irwin et al. (1995). To determine drift rates, empirical relationships such as Eqs. (C7-2) and (C7-3) are used. Comprehensive meteorological records are input in digital form, and empirical relationships to describe heat transfer, melting, and water retention in the snow pack are used. Typically, snow load patterns on the roof are simulated on an hour-by-hour basis for 30 to 40 winters, from which statistical analyses may be conducted to determine design load conditions, including uniform types of load distribution and unbalanced loads. Comparisons with full scale indicate that neglecting suspension is a conservative approach (Irwin 1994).

The Time Step B simulation method has both the benefits and liabilities of the Particle method. It does not require measurement of velocities over the roof surface or definition of empirical drift rates, as in the Time Step A method, but uses the combined falling and drifting results of the Particle method to define the snowfall/drift distribution determined from a physical model test to assign new snow depths each hour during the storm. In other respects, the model is very similar to the Time Step A method.

Chapter C8
ACCURACY, PRECISION, AND QUALITY ASSURANCE

C8.1 GENERAL

Wind is, by its very nature, a random process. This means that it is not possible to predict future values of wind speeds or wind effects from present or past values except in terms of statistical representations, including mean values; extreme values such as peaks, variance, or root-mean-square values; power spectral density functions; probability density functions; correlation functions; and various derived characteristics describing these representations. Consequently, almost all wind tunnel model studies involving simulated natural wind conditions require measurement of random fluctuating quantities such as wind speeds, pressures, forces, or motion.

Precision is the repeatability of a measurement without regard to how close the measurement is to the true value. *Accuracy* is the closeness of a measurement to the true value. It is possible for a highly precise measurement to have poor accuracy; for example, the use of a meter stick that is biased by being 20% longer than a true meter may give measurements that are repeatable within 1 mm but are accurate only to 200 mm. Acceptable accuracy is usually obtained by a measurement process that combines both a reasonable level of precision and systematic bias, or accuracy. Precision and bias must be balanced. A large expenditure to obtain one without the other may not yield an acceptable result. *Quality assurance* is the process by which a laboratory ensures that its accuracy and precision goals are routinely met.

Comparisons of full-scale and model data for a broad range of wind effects can be found in the literature. These include overall responses of wind-sensitive structures, including tall buildings, masts, towers, long-span bridges, and various specialty structures; local wind loads on components of both tall and low-rise buildings; and wind conditions at pedestrian level. The picture that has emerged is an encouraging one and confirms that properly scaled wind tunnel tests are capable of providing representative information on the action of wind on buildings and structures that meets reasonable accuracy requirements.

Many types of instruments are available for measuring quantities of interest in wind tunnel model studies. It is not the intent of this section to catalog these instruments along with their strengths and weaknesses; it deals, instead, with general issues of precision, accuracy, and quality assurance, which are important to the measurement of random variables.

In some situations, useful information may be obtained with more traditional techniques, including smoke-flow visualization and tufts. Flow visualization in turbulent flow shall be performed with due attention to proper photographic techniques for evaluation of the overall flow regime and for providing qualitative information on the potential presence of particular aerodynamic loading mechanisms.

Accurate measurement of random variables requires both precise quantification of the variable and its monitoring for a sufficient amount of time to establish stable estimates. Bendat and Piersol (1986) provide an excellent guide to the measurement and analysis of random data. Sections C8.2 and C8.3 draw from this reference as well as from practical experience in the measurement of wind loads and wind effects. The discussion centers on both questions of precision (repeatability) and accuracy (closeness to the true value) of the measurements.

C8.2 PRECISION

Wind tunnel tests usually involve scale models that are much smaller than the prototype and are tested at wind speeds that are lower than design values. For example, if an aeroelastic model is constructed with Froude number scaling, the properly scaled wind velocity is lower than the prototype speed by the square root of the geometric scale. For models where Froude number scaling is not employed, the test velocity is generally selected large enough to ensure that Reynolds number effects are either not important or can be accounted for in model design or analysis of results. Once this issue is dealt with, the actual speed selected represents a tradeoff between increased signal level (higher speed leading to larger pressures or forces) to obtain an improved electronic signal-to-noise ratio, and reduced frequency requirements (at lower speeds, fluctuations occur at a slower rate and frequency response requirements of the instrumentation are reduced).

A basic relationship used throughout wind engineering studies to relate wind speeds, model scales, and time or frequencies is

$$\frac{n_m L_m}{U_m} = \frac{n_p L_p}{U_p} \qquad \text{(C8-1)}$$

where n = frequency, L = a characteristic dimension, U = wind velocity, and subscripts m and p refer to model and prototype, respectively.

Rearranging terms produces the relationship between model frequencies or time, t, as

$$\frac{n_m}{n_p} = \frac{t_p}{t_m} = \frac{L_p}{L_m}\frac{U_m}{U_p} \qquad \text{(C8-2)}$$

These relationships allow the wind tunnel operator to judiciously select the appropriate wind tunnel test velocity or instrumentation frequency response characteristics for a given model scale and prototype wind conditions. In most cases, the frequency response characteristics of the instrumentation are restrained by the physical or electrical characteristics of the instrumentation system used. For example, it is well known that the frequency response characteristics of common tubing systems used for pressure measurements are dependent on the length and diameter of the tubing, changes in the diameter of the tubing, and the volume of the pressure transducer. For some length/diameter combinations of tubing, restrictions or equalizers are used to compensate for the effect of the tubing system on the dynamic pressures (the tubing system transfer function); refer to Bergh and Tijdeman (1967), Irwin et al. (1979), and Holmes (1987).

In order for a successful experiment to be performed, the wind tunnel, instrumentation, and data acquisition system must first be compatible, providing, for example, acceptable instrument electrical signal-to-noise at the speeds provided by the wind tunnel. In addition, frequency response of the instruments must be adequate to respond to the fluctuations induced by the modeled wind. If these and other requirements (which cannot all be specified here) are compatible, then the experiment design can proceed.

Eq. (C8-2), the size and speed range of the wind tunnel, the physical and electrical characteristics of the instrumentation, and the characteristics of the data acquisition system are all used to design an experiment with acceptable precision and accuracy. The design must consider such factors as the length, time, and frequency scaling based on Eq. (C8-2); Reynolds number independence; frequency content, amplitude, and noise level of the measured signal; frequency response, sensitivity, stability, dynamic range, and noise characteristics of the instrumentation; digital sample rate, dynamic range, sample time, aliasing possibilities, and channel-to-channel time lag of the data acquisition system; and the statistical character of the measured signal and the desired statistics of the output.

A number of references provide good discussions of proper experimental methods. For example, Bendat and Piersol (1986) and Durgin (1982) provide guides to the measurement and analysis of random data. Experimental methods specific to wind tunnel testing for wind loads can frequently be found in the *Journal of Wind Engineering and Industrial Aerodynamics*. Additional discussions of these issues are contained in ASCE MOP 67, *Wind Tunnel Studies of Buildings and Structures* (ASCE 1999).

Precision is only one factor affecting the overall accuracy and repeatability. The need for precise measurements should be balanced against the inherent variability in measuring random processes where, for example, large peak pressure coefficients can vary by 30% from one realization of the test to another due to the inherent physics of the flow.

C8.3 ACCURACY

From earlier discussion, it should be clear that the overall accuracy of the wind tunnel measurements depends on the precision of the measurements, the random characteristics of the quantity being measured, and the type of statistical quantity being determined. The first important component of accuracy is ensuring that the measuring instruments are calibrated.

Calibration is the comparison of a measurement by a laboratory instrument to that of a traceable instrument. That is, the measurement instrument has been calibrated against another instrument which has itself been calibrated. *Traceable calibration* means that there is a trail of calibrations leading eventually to comparison with a recognized national standard, such as those maintained by the U.S. National Institute of Standards and Technology (NIST). All laboratory measurements should be made by instruments with traceable calibrations. Calibration histories should be maintained by the laboratory.

The time interval over which data are gathered and the method of analysis are frequently important to the overall accuracy of the measurement in a turbulent wind environment. Bendat and Piersol (1986), Snyder (1981), and references in Snyder (1981) provide discussions of some of the considerations involved in selecting an appropriate averaging time and associated targets for accuracy. Mean values generally require the shortest amount of time to establish stable estimates; root-mean-square values and variances require somewhat longer times to obtain stable estimates than the mean. Extreme values or peaks, higher-order correlations, including spectral analysis, correlation functions, and coherence functions, may require lengthy sampling times and special analysis procedures to establish stable estimates.

As an example of time requirements, in areas of flow separation where large fluctuations are observed, the extreme values obtained from a single measurement of the maximum or minimum peak pressure may exhibit considerable variability for times corresponding to 1 to 2 h full scale. Most laboratories have instituted probabilistic methods for obtaining more stable estimates of extreme values by measuring a number of peaks and using analysis techniques. Several essentially equivalent methods exist, including Melbourne (1977), Peterka (1983), and Rofail and Kwok (1992).

The laboratory operator and end user should both be aware that increased accuracy requires increased testing time, and that the increases can be significant for some types of measurements and depend on the speeds used in the model tests. The target levels of precision and accuracy should be established with a realistic appreciation for the budget implications and consideration of all the other sources of uncertainty, including the wind climate. Because uncertainties are generally independent and the combined uncertainty is estimated from the square root of the sum of the squares of the individual values, it may be a waste of resources to double the sampling time in order to reduce the uncertainty in the peak pressure coefficient from, say, 10% to 7% if the uncertainty in the wind climate is 15%. Assuming that these are the only two uncertainties involved, reducing the uncertainty in the peak pressure coefficient from 10% to 7% would only reduce the total uncertainty from 18% to 17%.

This discussion is not intended to suggest that accuracy should not be a major focus in the laboratory study. Rather, it is intended to reinforce the understanding that the study of wind effects on structures involves the management of many potential sources of uncertainty, some of which would be common to both physical models and analytical or numerical studies. Thus, the operator and client should not develop unrealistic expectations for overall accuracy based on close attention to one or two parameters when a balanced overall treatment of all the sources of uncertainties would produce more reliable predictions.

C8.4 QUALITY ASSURANCE

The purpose of quality control of wind tunnel testing and wind engineering analysis is to ensure that they are carried out to satisfactory standards and that the results are reliable and consistent with the expectations of the sponsor.

In particular, the objectives of quality control are to ensure that:

- The scaling and representation of the properties of the natural wind conform as far as possible to the full scale; where they do not conform, it should be noted and understood;
- The scaling and representation of the models and physical processes affected by the wind conform to the full scale and perform as intended;
- Instruments used in experiments function correctly and consistently and are calibrated accurately; that computer programs for data acquisition, processing, editing, and analysis perform reliably and accurately;
- Theoretical interpretations suitably reflect the realities of the problem and the limitations and assumptions are stated and understood;

- As a long-term industry objective, full-scale verification of all key assumptions should be sought;
- The likelihood of human error is minimized in all aspects of the study; and
- Reporting is complete and understood by the sponsor.

The following outlines some practical procedures for achieving these objectives. It should be recognized that the procedures used in wind engineering are not absolute in character and are steadily evolving and expanding.

C8.5 CONDUCT OF TESTS AND ANALYSES

It is recommended that in all phases of wind tunnel testing and wind engineering analysis, careful project and facility records should be kept. These should include:

- Dates and times;
- The persons responsible for defining the tasks to be performed, setting up the experiments, running programs, checking installations, scheduling and controlling documentation;
- Wind tunnel condition and servicing, cleaning of screens, etc.;
- The identification and condition of equipment, calibration of instruments, gain settings, etc.;
- Operating schedule, interruptions;
- Computing software used (including identification of the particular version), file names of data; and
- Archiving of all results and data.

All errors encountered should be discussed openly (avoiding recriminations), investigated fully, and steps taken to avoid their recurrence. This process should be fully recorded. Good communications should be emphasized.

C8.6 SOURCES OF ERROR

Human errors are the most difficult to avoid. They are influenced by attitudes to work, the sense of professional responsibility, organization, and experience. The following approaches can be helpful in reducing and eliminating human and other errors.

- Assign responsibility for checking the quality. In some instances, it is desirable to assign a Quality Officer who is not otherwise involved in the test.
- Establish clear responsibilities for the tasks with definition of checks to be performed and signed off.
- Ensure that instrumentation and software used has been properly checked and commissioned. If software is modified, ensure that it is checked and modifications recorded.
- Clear and frank discussion of errors when they occur should be encouraged.
- Have work checked by a second person to the extent feasible.
- Perform periodic quality audits, in which the Quality Officer checks that all quality assurance procedures are being followed.
- Standardize experiments and computations.

The following techniques can be useful in identifying human and other errors: repeat measurements; dual instrumentation for critical measurements; graphic display (for example, trends in the variation of pressure coefficients with azimuth can identify problems); interdependency and redundancy (for example, integration of mean pressures on a tall building should be consistent with direct measurements of base shear or moment); comparison with theory; and checking of pressure tubing for location and leaks.

REFERENCES

Anno, Y. (1984). "Requirements for modelling of a snowdrift." *Cold Reg. Sci. and Tech.*, 8, 241–252.

ASCE. (1994). "Minimum design loads for buildings and other structures." *ASCE 7–93*, Reston, VA.

ASCE. (1996). "Minimum design loads for buildings and other structures." *ASCE 7–95*, Reston, VA.

ASCE. (1999). "Wind tunnel studies of buildings and structures." *ASCE Manuals and Reports on Engineering Practice No. 67*, Reston, VA.

ASCE. (2000). "Minimum design loads for buildings and other structures." *ASCE 7–98*, Reston, VA.

ASCE. (2003). "Minimum design loads for buildings and other structures." *ASCE 7–02*, Reston, VA.

American Society of Mechanical Engineers (ASME). (1986). "Quality assurance program requirements for nuclear facilities." *ANSI/ASME NQA-1–1986*, New York.

Bagnold, R. A. (1941). *The physics of blown sand and desert dunes*, Methuen, London.

Batts, M. E., Cordes, M. R., Russell, C. R., Shaver, J. R., and Simiu, E. (1980). "Hurricane windspeeds in the United States." *Rep. No. BSS-124*, National Bureau of Standards, U.S. Dept. of Commerce, Washington, DC.

Bendat, J. S., and Piersol, A. G. (1986). *Random data analysis and measurement procedures*, 2nd Ed., John Wiley & Sons, New York.

Bergh, H., and Tijdeman, H. (1967). "Theoretical and experimental results for the dynamic response of pressure measuring systems." *Rep. NLR-TR F.238*, National Aero- and Astronautical Research Institute, Amsterdam.

Boggs, D. W., and Peterka, J. A. (1989). "Aerodynamic model tests on tall buildings." *J. Eng. Mech.*, 115(3), 618–635.

Browning, K. A. (1964). "Airflow and precipitation trajectories within severe local storms which travel to the right of the winds." *J. Atmos. Sci.*, 21, 634–639.

Busch, N. E. (1973). "On the mechanics of atmospheric turbulence." *Workshop on micrometeorology*, American Meteorological Society, Boston, MA.

Canadian Standards Association (CSA). (1991). *Quality assurance program: Category 1, CAN3-Z299.1–8b*, Mississauga, Ontario, Canada.

Cermak, J. E. (1975). "Applications of fluid mechanics to wind engineering: A Freeman Scholar lecture." *Trans. ASME, J. Fluids Eng., Ser. 1*, 97(1), 9–38.

Cermak, J. E. (1981). "Wind tunnel design for physical modeling of atmospheric boundary layers." *J. Eng. Mech. Div.*, 107(3), 523–642.

Cermak, J. E. (1982). "Physical modeling of the atmospheric boundary layer (ABL) in long boundary-layer wind tunnels (BLWT)." *Proc., Int. Workshop on Wind Tunnel Modeling Criteria and Techniques in Civil Engineering Applications*, Gaithersburg, MD, T. A. Reinhold, ed., Cambridge University Press, New York, 97–137.

Cermak, J. E. (1984). "Physical modeling of flow and dispersion over complex terrain." *Boundary Layer Meteorol.*, 30, 261–292.

Cermak, J. E., Cochran, L. S., and Leffler, R. D. (1993). "Wind-tunnel modelling of the atmospheric surface layer." *Proc., 3rd Asia-Pacific Symp. on Wind Engineering*, Department of Civil and Structural Engineering, University of Hong Kong, Hong Kong, December 13–25.

Chen, X., and Kareem, A. (2004). "Efficacy of the implied approximation in the identification of flutter derivatives." *J. Struct. Eng.*, 130(12), 2070–2074.

Cook, N. J. (1982). "Towards better estimation of extreme winds." *J. Wind Eng. Ind. Aerodyn.*, 9, 295–323.

Cook, N. J. (1985). *Designers guide to wind loading of building structures, Part I: Background, damage survey, wind data, and structural classification*, Butterworths, London.

Counihan, J. (1969). "An improved method of simulating an atmospheric boundary layer in a wind tunnel." *Atmos. Env.*, 3, 197–214.

Counihan, J. (1975). "Adiabatic atmospheric boundary layers: A review and analysis of data from the period 1880–1972." *Atmos. Env.*, 9, 871–905.

Davenport, A. G. (1960). "Rationale for determining design wind velocities." *J. Struct. Div.*, 86, 39–68.

Davenport, A. G. (1961). "The spectrum and horizontal gustiness near the ground in high winds." *Quarterly J. Royal Meteorol. Soc.*, 87, 194–221.

Davenport, A. G. (1968). "The dependence of wind loading on meteorological parameters." *Proc., Int. Res. Sem. on Wind Effects on Buildings and Structures*, National Research Council of Canada, University of Toronto Press, Toronto.

Davenport, A. G. (1977). "The prediction of risk under wind loading." *Proc., 2nd Int. Conf. on Structural Safety and Reliability*, Munich, International Association for Structural Safety and Reliability, New York.

Davenport, A. G. (1982). "Consideration in relating wind tunnel results to design for a specific site." *Proc., Int. Workshop on Wind Tunnel Modeling Criteria and Techniques in Civil Engineering Applications*, Gaithersburg, MD, T. A. Reinhold, ed., Cambridge University Press, New York.

Davenport, A. G. (1983). "The relationship of reliability to wind loading." *J. Wind Eng. Ind. Aerodyn.*, 13, 3–27.

Davenport, A. G., and Isyumov, N. (1967). "The application of the boundary layer wind tunnel to the prediction of wind loading." *Proc., Int. Res. Sem. on Wind Effects on Buildings and Structures, 2nd Int. Conf. on Wind Engineering*, Ottawa, Canada, 201–230.

Davenport, A. G., Isyumov, N., King, J. P. C., Novak, M., Surry, D., and Vickery, B. J. (1985). "BLWT II: The design and performance of a new boundary layer wind tunnel." *Proc., 5th U.S. Nat. Conf. on Wind Engineering*, Texas Tech University Press, Lubbock, TX.

Davenport, A. G., King, J. P. C., and Larose, G. L. (1992). "Taut strip model tests." *Proc., 1st Int. Symp. on the Aerodynamics of Large Bridges*, Copenhagen, Denmark, February 19–21, A. A. Balkema, Rotterdam, 113–124.

Durgin, F. (1982). "Instrumentation requirements for aerodynamic pressure and force measurements on buildings and structures." *Wind tunnel modeling for civil engineering applications*, Cambridge University Press, New York, 329–348.

Durst, C. (1960). "Wind speeds over short periods of time." *The Meteorological Magazine*, 89(1056), 181–187.

Dyunin, A. K. (1963). "Solid flux of snow bearing airflow." *Tech. Trans. TT-1102* (from Russian), National Research Council of Canada, Ottawa.

Etkin, B. (1971). "The interaction of precipitation with complex flows." *Proc., 3rd Int. Conf. on Wind Effects on Buildings and Structures*, Tokyo, Saikon Shuppan Co. Ltd., Tokyo, 135–143.

ESDU International (ESDU). (1993). "Characteristics of atmospheric turbulence near the ground." *ESDU 85020*, ESDU International, PLC, London.

Gamble, S. L., Kochanski, W. K., and Irwin, P. A. (1991). "Finite area element snow loading prediction: Applications and advancements." *Proc., 8th Int. Conf. on Wind Engineering*, Elsevier, London, Ontario.

Gamble, S. L., Miltenburg, R. J., Cicci, M. D., and Accardo, M. (2001). "Prediction of local exterior pressures from wind tunnel studies using a time history analysis approach." *Proc., Americas Conf. on Wind Engineering*, Clemson, NC, American Association for Wind Engineering, Williamsburg, VA.

Georgiou, P. N. (1985). "Design windspeeds in tropical cyclone-prone regions." Doctoral dissertation, Faculty of Engineering Science, University of Western Ontario, London, Ontario.

Georgiou, P. N., Davenport, A. G., and Vickery, B. J. (1983). "Design wind speeds in regions dominated by tropical cyclones." *Proc., 6th Int. Conf. on Wind Engineering*, Gold Coast, Australia, March 21–25; Auckland, New Zealand, April 6–7.

Golden, J. H., and Snow, J. T. (1991). "Mitigation against extreme windstorms." *Rev. of Geophys.*, 29(4), 477–504.

Gomes, L., and Vickery, B. J. (1976). "On the prediction of tropical cyclone gust speeds along the northern Australian coast." *Inst. Eng. Aust., C.E. Trans.*, 18(2), 40–49.

Gomes, L., and Vickery, B. J. (1977). "On the prediction of extreme wind speeds from the parent distribution." *J. Ind. Aerodyn.*, 2(1), 21–36.

Gumley, S. A. (1983). "Detailed design method for pneumatic tubing systems." *Proc. 6th Int. Conf. on Wind Engineering*, Gold Coast, Australia.

Harris, R. I., and Deaves, D. M. (1980). "The structure of strong winds." *Paper 4, CIRA Conf. on Wind Engineering in the Eighties*, Construction Industry Research and Information Association, London.

Hellman, G. (1916). "Über die bewegung der luft in den untersten schichten der atmosphäre." *Meteorol. Z.*, 34, 273.

Ho, T. C. E., Lythe, G. R., and Isyumov, N. (1999). "Structural loads and responses from the integration of instantaneous pressures." *Proc., 10th Int. Conf. on Wind Engineering*, A. A. Balkema, Rotterdam, 1505–1510.

Holmes, J. D. (1982). "Techniques and modeling criteria for the measurement of external and internal pressures." *Wind Tunnel Modeling for Civil Engineering Applications*, T. A. Reinhold, ed., Cambridge University Press, New York.

Holmes, J. D. (1986). "The application of probability theory to the directional effects of wind." *Proc., 10th Australasian Conf. on the Mechanics of Structures and Materials*, University of Adelaide, Adelaide SA, Australia.

Holmes, J. D. (1987). "Mode shape corrections for dynamic response to wind." *Eng. Struct.*, 9, 210–212.

Holmes, J.D., and Lewis, R. E. (1987). "Optimization of dynamic-pressure measurement systems. I. Single point measurements." *J. Wind Eng. Ind. Aerodyn.*, 25, 249–273.

Holton, J. R. (1971). *An introduction to dynamic meteorology*, Academic Press, New York.

Irwin, H. P. A. H., Cooper, K. R., and Girard, R. (1979). "Correction of distortion effects caused by tubing systems in measurements of fluctuating pressures." *J. Ind. Aerodyn.*, 5, 93–107.

Irwin, P. A. (1979a). "Centre of rotation for torsional vibration of bridges." *J. Ind. Aerodyn.*, 4, (2), 123–132.

Irwin, P. A. (1979b). "Cross-spectra of turbulence in isotropic turbulence." *Boundary Layer Meteorol.*, 16, 237–243.

Irwin, P. A. (1982). "Model studies of the dynamic response of tall buildings to wind." *Proc., Canadian Soc. of Civil Engineering, Annual Conference*, Edmonton, May 27–28, 1, 285–302.

Irwin, P. A. (1987). "Pressure model techniques for cladding loads." *Proc., 7th Int. Conf. on Wind Engineering*, Aachen, Germany.

Irwin, P. A. (1992). "Full aeroelastic model tests." *Proc., 1st Int. Symp. on the Aerodynamics of Large Bridges*, Copenhagen, Denmark, February 19–21, A. A. Balkema, Rotterdam, 125–135.

Irwin, P.A. (1994). "Hybrid physical/computer modelling of snow drifting." *Special ASCE Publication on the Modeling of Windblown Snow and Sand, ASCE/ISSW Snow Science Workshop*, Snowbird, Utah, October 31.

Irwin, P. A., and Gamble, S. L. (1988). "Prediction of snow loading on the Toronto SkyDome." *Proc., 1st Int. Conf. on Snow Engineering*, Santa Barbara, CA, CRREL, Hanover, NH.

Irwin, P. A., and Gamble, S. L. (1992). "Model and computer studies of snow loading." *Proc., Int. Cong. on Innovative Large Span Structures*, International Association of Shell and Space Structures, and Canadian Society of Civil Engineers, Toronto, 381–392.

Irwin, P. A., Gamble, S. L., and Taylor, D. A. (1995). "Effects of roof size and heat transfer on snow load: Studies for the 1995 NBC." *Canadian J. Civ. Eng*, 22, 770–784.

Irwin, P. A., and Kochanski, W. W. (1995). "Measurement of structural wind loads using the high frequency pressure integration method." *Proc., ASCE Structures Congress XIII*, Boston, April 2–5, 1631–1634.

Irwin, P. A., and Wardlaw, R. L. (1979). "A wind tunnel investigation of a retractable fabric roof for the Montreal Olympic stadium." *Proc. 5th Int. Conf. on Wind Engineering*, Fort Collins, CO, July, Pergamon Press, New York, 925–938.

Irwin, P. A., and Williams, C. J. (1983). "Application of snow simulation model tests to planning and design." *Proc., Eastern Snow Conf. 28, 40th Annual Meeting*, Toronto, Eastern Snow Conf., Peterborough, Ontario, 118–130.

Isyumov, N. (1971). "An approach to the prediction of snow loads." Doctoral dissertation, University of Western Ontario, London, Ontario.

Isyumov, N. (1982). "The aeroelastic testing of tall buildings." *Proc., Int. Workshop on Wind Tunnel Modeling Criteria and Techniques in Civil Engineering Applications*, Gaithersburg, MD, April, Cambridge University Press, New York, 373–407.

Isyumov, N., Case, P. C., Ho, T. C. E., and Soegiarso, R. (2001). "Wind tunnel model studies to predict the action of wind on

the projected 558-m Jakarta Tower." *Wind & Structures,* 4(4), 299–314.
Isyumov, N., and Mikitiuk, M. (1992). "Wind tunnel modeling of snow accumulations on large area roofs." *Proc., 2nd Int. Conf. on Snow Engineering*, Santa Barbara, CA, CRREL, Hanover, NH, 181–193, 1992.
Isyumov, N., Mikitiuk, M. J., Case, P. C., Lythe, G. R., and Welburn, A. (2003). "Predictions of wind loads and responses from simulated tropical storm passages." *Proc., 11th Int. Conf. on Wind Engineering*, D. Smith, ed., Texas Tech University Press, Lubbock, TX, 254–260.
Iversen, J. D. (1982). "Small-scale modelling of snowdrift phenomena." *Proc., Int. Workshop on Wind Tunnel Modelling Criteria in Civil Engineering Applications*, Gaithersburg, MD, Cambridge University Press, New York, 522–545.
Jensen, M. (1958). "The model-law for phenomena in natural wind." *IngeniΩren*, 2, 121–128.
Kaimal, J. C., and Finnigan, J. J. (1994). *Atmospheric boundary layer flows: Their structure and measurement,* Oxford University Press, New York.
Kaimal, J. C., Wyngaard, J. C., Izumi, Y., and Coté, O. R. (1972). "Spectral characteristics of surface-layer turbulence." *Quarterly J. Royal Meteorol. Soc.*, 98, 563–589.
Kareem, A. (1982). "Fluctuating wind loads on buildings." *J. Eng. Mech.*, 6, 1086–1102.
Kareem, A., Lu, P. C., Finnigan, T. D., and Liu, S-L. V. (1987). "Aerodynamic loads on a typical tension leg platform." *Ocean Eng.*, 14(3), 201–231.
Kawakita, S., Bienkiewicz, B., and Cermak, J. E. (1992). "Aeroelastic model study of suspended cable roof." *J. Wind Eng. Ind. Aerodyn.*, 42, 1459–1470.
Kind, R. J. (1981). "Snow drifting." Chapter 8, *Handbook of snow: Principles, processes, management and use,* D. M. Gray and D. H. Male, eds., Pergamon Press, New York.
Kind, R. J. (1982). "Aeroelastic modeling of membrane structures." *Proc., Int. Workshop on Wind Tunnel Modeling Criteria and Techniques in Civil Engineering Applications*, Gaithersburg, MD, Cambridge University Press, New York, 429–439.
Kind, R. J. (1986). "Snowdrifting: A review of modelling methods." *Cold Reg. Sci. and Tech.*, 12, 217–228.
Kind, R. J., and Murray, S. B. (1982). "Saltation flow measurements relating to the modelling of snow drifting." *J. Wind Eng. Ind. Aerodyn.*, 10, 89–102.
Kobayashi, D. (1973). "Studies of snow transport in low level drifting snow." *Rep. No. A31,* Institute of Low Temperature Science, Sapporo, Japan, 1–58.
Kwok, K. C. S., Kim, D. H., Smedley, D. J., and Rohde, H. F. (1991). "Snowdrift around buildings for Antarctic environment." *Proc., 8th Int. Conf. on Wind Engineering*, Elsevier, London, Ontario, 2797–2808.
Lepage, M. R., and Irwin, P. A. (1985). "A technique for combining historical wind data with wind tunnel tests to predict extreme wind loads." *Proc., 5th U.S. Nat. Conf. of Wind Engineering*, Texas Tech University Press, Lubbock, TX, 2B-71-2B-78.
Lettau, H. H. (1962). "Theoretical wind spirals in the boundary layer of a barotropic atmosphere." *Beiträge zur Atmosphärenphysik*, 35, 195–212.
Melbourne, W. H. (1977). "Probability distributions associated with the wind loading on structures." *IEAust., Civ. Eng. Trans.*, CE19, 59–67.
Mellor, M. (1965). "Blowing snow." *Cold regions science and engineering, Part III, Section A3C*, U.S. Army Cold Regions Research and Engineering Laboratory, Hanover, NH.
Meroney, R. N. (1980). "Wind-tunnel simulation of the flow over hills and complex terrain." *J. Wind Eng. Ind. Aerodyn.*, 5, 297–321.
Miyata, T., Yokoyama, K., Yasuda, M., and Hikami, Y. (1992). "Akashi Kaikyo Bridge: Wind effects and full model wind tunnel tests." *Proc., 1st Int. Symp. on the Aerodynamics of Large Bridges*, Copenhagen, Denmark, February 19–21, A. A. Balkema, Rotterdam, 217–236.
Novak, M., and Kassem, M. (1990). "Effect of leakage and acoustical damping of free vibration of light roofs backed by cavities." *J. Wind Eng. Ind. Aerodyn.*, 36, 289–300.
Odar, F. (1965). "Simulation of drifting snow." *Research Rep. 174*, Cold Regions Research and Engineering Laboratory, Hanover, NH.
Olesen, H. R., Larsen, S. E., and H jstrup, J. (1984). "Modelling velocity spectra in the lower part of the planetary boundary layer." *Boundary-Layer Meteorol.*, 29, 285–312.
O'Rourke, M., and Weitman, N. (1992). "Laboratory studies of snow drifts on multilevel roofs." *Proc., 2nd Int. Conf. on Snow Engineering*, Santa Barbara, CA, CRREL, Hanover, NH, 195–206.
Panofsky, H. A., and Dutton, J. A. (1983). *Atmospheric turbulence: Models and methods for engineering applications,* John Wiley & Sons, New York.
Peterka, J. (1983). "Selection of local peak pressure coefficients for wind tunnel studies of buildings." *J. Wind Eng. Ind. Aerodyn.,* 13, 477–488.
Peterka, J. (1992). "Improved extreme wind predictions in the United States." *J. Wind Eng. Ind. Aerodyn.*, 41, 533–541.
Peterka, J., and Shahid, S. (1998). "Design gust wind speeds for the U.S." *J. Struct. Eng.*, 124, 207–214.
Peterka, J., and Petersen, R. L. (1990). "On the relaxation of saltation length as a modeling criteria for particulate transport." *J. Wind Eng. Ind. Aerodyn.*, 36, 867–876.
Peterka, J. A., and Esterday, W. S. (2004). "Roof design snow loads by wind tunnel test and analysis." *Proc., ASCE Structures Cong. 2004*, Nashville, TN.
Petersen, R., and Cermak, J. (1988). "Application of physical modeling for assessment of snow loading and drifting." *Proc., 1st Int. Conf. on Snow Engineering*, Santa Barbara, CA, CRREL, Hanover NH, 276–285.
Reinhold, T. A., and Vickery, P. J. (1990). "Variable mode shape high-frequency force balance." *NSF Contract ISI 8660406*, National Science Foundation, Washington, DC.
Rofail, A. W., and Kwok, C. W. (1992). "Reliability study of wind tunnel results for cladding pressures." *J. Wind Eng. Ind. Aerodyn.*, 42, 2413–2424.
Roselas, M. (1983). "A novel technique for measuring spatially-averaged pressure: Its development and application to low-rise buildings." Master's thesis, University of Western Ontario, London, Ontario.
Russell, L. R. (1971). "Probability distributions for hurricane effects." *J. Waterways, Harbors, and Coastal Eng. Div.*, 97(WW1), 139–154.
Russell, L. R., and Schueller, G. F. (1974). "Probabilistic models for Texas Gulf coast hurricane occurrences." *J. Petroleum Tech.*, March, 279–288.
Scanlan, R. H. (1992). "Wind dynamics of long span bridges." *Proc. 1st Int. Symp. on the Aerodynamics of Large Bridges*, Copenhagen, Denmark, February 19–21, A. A. Balkema, Rotterdam, 47–57.
Scanlan, R. H., and Tomko, J. J. (1971). "Airfoil and bridge deck flutter derivatives." *J. Eng. Mech. Div.*, 97(6), 1717–1737.
Schlichting, H. (1979). *Boundary layer theory,* McGraw-Hill, New York.

Schmidt, R.A. (1986). "Transport rate of drifting snow and the mean wind speed profile." *Boundary-Layer Meteorol.*, 34, 213–241.

Simiu, E. (1974). "Wind spectra and dynamic alongwind response." *J. Struct. Div.*, 100, 1877–1910.

Simiu, E., Changery, M. J., and Filliben, J. J. (1979). "Extreme windspeed of 129 stations in the contiguous United States." *NBS Building Science Series 118*, U.S. Dept. of Commerce, Washington, DC.

Simiu, E., Filliben, J. J., and Shaver, J. R. (1982). "Short-term records and extreme wind speeds." *J. Struct. Div.*, 108(11), 2571–2577.

Simiu, E., and Heckert, N. A. (1996). "Extreme wind distribution tails: A 'peaks over threshold' approach." *J. Struct. Eng.*, 122(11), 539–547.

Simiu, E., and Scanlan, R. H. (1996). *Wind effects on structures: An introduction to wind engineering*, 3rd Ed., John Wiley & Sons, New York.

Snyder, W. H. (1981). "Guidelines for fluid modeling of atmospheric diffusion." *Rep. EPA-600/8–81–009*, U.S. Environmental Protection Agency, Washington, DC.

Standen, N. M. (1972). "A spire array for generating thick turbulent shear layers for natural wind simulation in wind tunnels." *Tech. Rept. LTR-LA-94*, National Aeronautical Establishment, Ottawa, Canada.

Stathopoulos, T. (1982). "Techniques and modeling criteria for measuring area-averaged pressures." *Wind tunnel modeling for civil engineering applications*, Cambridge University Press, New York.

Stathopoulos, T., Surry, D., and Davenport, A. G. (1979). "Internal pressure characteristics of low-rise buildings due to wind action, *Proc., 5th Int. Conf. on Wind Engineering,* Fort Collins, CO.

Steckley, A. C. (1989). "Motion-induced wind forces on chimneys and tall buildings." Doctoral dissertation, University of Western Ontario, London, Ontario.

Steckley, A. C., Vickery, B. J., and Isyumov, N. (1992). "The use of integrated pressures to determine overall wind-induced responses." *J. Wind Eng. Ind. Aerodyn.*, 41–44, 1023–1034.

Surry, D., and Stathopoulos, T. (1977). "An experimental approach to the economical measurement of spatially averaged wind loads." *J. Ind. Aerodyn.*, 2, 385–397.

Tabler, R. D. (1980). "The geometry and density of drifts formed by snow fences." *J. Glaciol.*, 26, 405–419.

Templin, J. T., and Cooper, K. R. (1980). "Design and performance of a multi-degree-of-freedom aeroelastic building model. *Proc., 4th Colloq. on Industrial Aerodynamics*, Aachen, Germany, June 18–20, Fachhochschule Aachen, Part 2, 123–135.

Templin, J. T., and Schriever, W. R. (1982). "Loads due to drifted snow." *J. Struct. Div*, 108(8), 1916–1925.

Tieleman, H. W. (1992). "Wind characteristics in the surface layer over heterogeneous terrain." *J. Wind Eng. Ind. Aerodyn.*, 41, 329–340.

Tieleman, H. W. (1995). "Universality of velocity spectra." *J. Wind Eng. Ind. Aerodyn.*, 56,55–70.

Toyoda, K., and Tomabechi, T. (1992). "Development of a wind tunnel for the study of snowdrifting." *Proc., 2nd Int. Conf. on Snow Engineering*, Santa Barbara, CA, CRREL, Hanover, NH, 207–214.

Tryggvason, B. V. (1979a). "Aeroelastic modelling of pneumatic and tensioned fabric structures." *Proc., 5th Int. Conf. on Wind Engineering*, Fort Collins, CO, Pergamon Press, New York, 1061–1072.

Tryggvason, B. V. (1979b). "Computer simulation of tropical cyclone wind effects for Australia." *Wind Engineering Rep. 2/79*, Dept. Civil and Systems Engineering, James Cook University, Townsville, Australia.

Tryggvason, B. V., Surry, D., and Davenport, A. G. (1976). "Predicting wind-induced response in hurricane zones." *J. Struct. Div.*, 102(12), 2333–2350.

Tschanz, T. (1982). "The base balance measurement technique and applications to dynamic wind loading of structures." Doctoral dissertation, University of Western Ontario, Faculty of Engineering Science, London, Ontario, Canada.

Twisdale, L. A. (1978). "Tornado data characterization and windspeed risk." *J. Struct. Div.*, 104(10), 1611–1630.

Twisdale, L. A., and Vickery, P. J. (1992). "Research on thunderstorm wind design parameters." *J. Wind Eng. Ind. Aerodyn.*, 41, 545–556.

Twisdale, L. A., and Vickery, P. J. (1993). "Analysis of thunderstorm occurrences and windspeed statistics." *Proc., 7th U.S. Nat. Conf. on Wind Engineering*, Irvine, CA, University of California–Los Angeles, Los Angeles.

Vickery, B. J. (1973). "On the use of balloon data to define wind speeds for tall structures." *Research Rep. R226*, School of Civil Engineering, University of Sydney, Australia.

Vickery, P. J., Steckley, A. C.., Isyumov, N., and Vickery, B. J. (1985). "The effect of mode shape on the wind induced response of tall buildings." *Proc., 5th U.S. Nat. Conf. on Wind Engineering*, Texas Tech University Press, Lubbock, TX, 1B-41–1B-48.

Vickery, P. J., and Twisdale, L. A. (1995a). "Prediction of hurricane windspeeds in the United States." *J. Struct. Eng.*, 11(121), 1691–1699.

Vickery, P. J., and Twisdale, L. A. (1995b). "Wind field and filling models for use in hurricane wind speed simulations." *J. Struct. Eng.*, 11(121), 1700–1709.

von Kármán, T. (1948). "Progress in the statistical theory of turbulence." *Proc., National Academy of Sciences*, Washington, DC, 530–539.

Whitbread, R. E. (1975). "The measurement of non-steady wind forces on small-scale building models." *Proc., 4th Int. Conf. on Wind Effects on Buildings and Structures*, Heathrow, England, Cambridge University Press, London, 369–379.

Wieringa, J. (1992). "Updating the Davenport roughness classification." *J. Wind Eng. Ind. Aerodyn.*, 42, 357–368.

Williams, C. J., Gamble, S. L., and Kochanski, W. (1992). "Snow load prediction in the Andes Mountains and downtown Toronto." *Proc., 2nd Int. Conf. on Snow Engineering*, Santa Barbara, CA, CRREL, Hanover, NH, 135–145.

Wuebben, J. L. (1978). "A hydraulic model investigation of snow drifting." *Rep. 78–16*, Cold Regions Research and Engineering Laboratory, U.S. Army Corps of Engineers, Hanover, NH.

Zhou, Y., and Kareem, A. (2003). "Aeroelastic balance." *J. Eng. Mech.*, 129(3), 282–292.

INDEX

Page numbers followed by f and t indicate figures and tables, respectively.